Astronomy
A Beginner's Guide

ONEWORLD BEGINNER'S GUIDES combine an original, inventive, and engaging approach with expert analysis on subjects ranging from art and history to religion and politics, and everything in-between. Innovative and affordable, books in the series are perfect for anyone curious about the way the world works and the big ideas of our time.

aesthetics
africa
american politics
anarchism
ancient philosophy
animal behaviour
anthropology
anti-capitalism
aquinas
archaeology
art
artificial intelligence
the baha'i faith
the beat generation
the bible
biodiversity
bioterror & biowarfare
the brain
british politics
the Buddha
cancer
censorship
christianity
civil liberties
classical music
climate change
cloning
the cold war
conservation
crimes against humanity
criminal psychology
critical thinking
the crusades
daoism
democracy
descartes
dewey
dyslexia
economics
energy
engineering
the english civil wars

the enlightenment
epistemology
ethics
the european union
evolution
evolutionary psychology
existentialism
fair trade
feminism
forensic science
french literature
the french revolution
genetics
global terrorism
hinduism
history
the history of medicine
history of science
homer
humanism
huxley
imperial china
international relations
iran
islamic philosophy
the islamic veil
jazz
jesus
journalism
judaism
justice
lacan
life in the universe
literary theory
machiavelli
mafia & organized crime
magic
marx
medieval philosophy
the middle east
modern slavery
NATO

the new testament
nietzsche
nineteenth-century art
the northern ireland conflict
nutrition
oil
opera
the palestine–israeli conflict
parapsychology
particle physics
paul
philosophy
philosophy of mind
philosophy of religion
philosophy of science
planet earth
populism
postmodernism
psychology
quantum physics
the qur'an
racism
rawls
reductionism
religion
renaissance art
the roman empire
the russian revolution
shakespeare
shi'i islam
the small arms trade
stalin
sufism
the torah
the united nations
the victorians
volcanoes
war
the world trade organization
world war II

Astronomy
A Beginner's Guide

William H. Waller

ONEWORLD

A Oneworld Book

First published by Oneworld Publications in 2022
Reprinted, 2023

Copyright © William H. Waller 2022

The moral right of William H. Waller to be identified as the author of this
work has been asserted by him in accordance with the Copyright, Designs,
and Patents Act 1988

All rights reserved
Copyright under Berne Convention
A CIP record for this title is available from the British Library

ISBN 978-0-86154-400-4
eISBN 978-0-86154-401-1

Every effort has been made to trace copyright holders for the use of material in
this book. The publisher apologizes for any errors or omissions herein and would
be grateful if they were notified of any corrections that should be incorporated in
future reprints or editions of this book.

Typeset by Geethik Technologies
Printed and bound in Great Britain by Clays Ltd, Elcograf S.p.A.

Oneworld Publications
10 Bloomsbury Street
London WC1B 3SR
England

Stay up to date with the latest books,
special offers, and exclusive content from
Oneworld with our newsletter

Sign up on our website
oneworld-publications.com

Dedicated to Paul W. Hodge (1934–2019)

Gentle mentor, colleague, and friend

Contents

Preface   1

Introduction   4

I Our place in space

1 First impressions   9

2 The day and night sky   15

3 Cosmic perspectives   40

II Constituents of the cosmos

4 Introducing our Solar System   79

5 The Sun: our star   95

6 Stars and planets beyond the Sun's
  domain   105

7 The Milky Way Galaxy   126

8 Menageries of galaxies and their cosmic
  expansion   150

III Our moment in time

9 The hot big bang   171

10 The emergence of galaxies   186

11 The birth of stars and planets   194

12 Cycles of life and death among the stars 203

13 Conundrums of matter and energy 211

14 Our cosmic inheritance 233

Recommended reading and resources 240

Index 243

Preface

I started writing this book while nestled high up in the control room of the Mayall four-meter telescope atop Kitt Peak National Observatory in southern Arizona. My observing run was turning out to be more terrestrial than celestial, with several rainy nights followed by a stubborn fog and topped off by a fierce windstorm. Fortunately for me, I was not alone in my nighttime vigils. A longtime colleague from my graduate student days and a rising graduate student had joined me. Already, the grad student had proved her superiority in working the computer interface, leaving the two of us veteran astronomers in the dust. As we fussed over our observing plan and shared our most colorful stories and YouTube videos, I could feel a direct connection with the many intrepid astronomers who had paved the way for us.

Four centuries before that ill-fated observing run, the Italian mathematician Galileo Galilei was pioneering the use of optical telescopes for astronomical studies. Using spyglasses of his own creation, Galileo perceived celestial wonders of profound consequence. From mountains on the Moon to spots on the Sun, apparitions of Venus that varied in phase and size, tiny moons in orbit around Jupiter, and the Milky Way resolved into countless stars, Galileo's observations irrevocably upended our Earthbound perspective of the Universe.

At around the same time, the German mathematician, astronomer, and astrologer Johannes Kepler was analyzing the finest naked-eye observations ever made of planetary positions and motions. Building on the heliocentric planetary system of Nicolaus Copernicus, Kepler had crafted a mathematical model whereby the six known planets (Mercury, Venus, Earth, Mars, Jupiter, and Saturn) traveled in circular orbits around the Sun. To Kepler, the relative spacing of these orbits was constrained by the mathematics of alternately nested polyhedrons and spheres of crystalline composition. He imagined that the rubbing of the polygons against the spheres would produce a music of divine import. Alas, his "music of the spheres" concept ran counter to his own careful analysis of the observed motion of Mars across the sky. Abandoning his exquisite model, Kepler let Mars travel in a slightly elliptical trajectory that sped up when nearest the Sun and slowed when farthest away. His resulting three laws of planetary motion – and Isaac Newton's subsequent explanation based on the concept of universal gravitation – epitomize how hard-won truths can emerge from the crucible of meticulous observations, imaginative analysis, and due diligence.

Through the dedication and genius of Galileo, Kepler, Newton, Herschel, Leavitt, Shapley, Payne-Gaposhkin, Hubble, and many other astronomers, we can today behold a truly wondrous Universe – as vast, rich, and transformative as one could ever imagine. The story continues, of course, with new discoveries often coming at breakneck speed. In writing this book, I have not tried to present an updated compendium of all that has been discovered. Instead, I have tried to provide a true beginner's guide to the content, structure, birth, and continuing evolution of the Universe. This has meant glossing over many worthwhile topics for the sake of maintaining some semblance of coherence. I have also interjected some personal reflections that may help to provide some human relief to what is otherwise a far-reaching cosmic narrative. If some of these ruminations betray me as a bit of a blowhard, so be it. Among astronomers, I would not be alone.

My own attempts to communicate have been aided by the kind generosity and empathetic support of myriad colleagues, friends, and family members. I am especially indebted to the scientists, faculty, staff, and students at the Harvard-Smithsonian

Center for Astrophysics, University of Washington, University of Massachusetts, Tufts University, and Rockport Public Schools who have helped to guide my learning and teaching of astronomy over the decades. My many colleagues within NASA's Space Science Education and Public Outreach "universe" have enabled me to bridge the gap between formal education and the more public forms of communication in astronomy. Amateur astronomers associated with the Gloucester Area Astronomy Club and other like-minded organizations have kept my own enthusiasm kindled with convivial gatherings and inspiring star parties. International colleagues have reminded me that all people share a fascination with the cosmos that deserves nurturing. The book itself greatly benefited from Leigh Slingluff, who skillfully crafted many of the figures. I am deeply grateful to Jonathan Bentley-Smith and his colleagues at Oneworld who expertly and patiently shepherded the manuscript into the book that you see today. Lastly, relatives far and wide, present and passed, have given me unwavering support to pursue my astronomical interests and science-writing projects such as this Beginner's Guide. These blessings, plus the ever-compelling wonders of the day and night sky, have made all the difference.

Bill Waller
Rockport, Massachusetts

Introduction

> *The Cosmos is all that there is or ever was or ever will be. Our fee-*
> *blest contemplations of the Cosmos stir us – there is a tingling in the*
> *spine, a catch in the voice, a faint sensation, as if a distant memory,*
> *of falling from a height. We know we are approaching the greatest of*
> *mysteries.*
>
> Carl Sagan, *Cosmos*

The field of astronomy encompasses the entire Universe and all
that it contains. Those who study astronomy routinely consider
the most wondrous phenomena on the grandest of scales. Yet,
to be a professional astronomer today is to be among the rar-
est of breeds. Worldwide, the International Astronomical Union
represents about 11,000 astronomers as members. An arbitrary
doubling of that figure would yield one astronomer for every
300,000 people on the planet – not quite one in a million but
certainly getting there. Despite these paltry numbers at the professional and graduate student level, astronomy is one of the most
– if not *the* most – popular of sciences. Around the world, hundreds of millions of people engage with the wonders of space via
academic courses, amateur clubs, magazines, planetarium shows,
television shows, websites, and books. This Beginner's Guide is
intended to serve the interests of curious readers who are looking for a digestible guide to astronomy and to the cosmos that
we all share.

Today, we find ourselves in a weird paradox. We have never
known so much about the Universe. Yet we have also never
known so little about the Universe. From discoveries of planets
around other stars, to images of newborn galaxies at the edge of

space and time, we have fathomed the deep as never before. What we have found has consistently surprised and challenged us. Each planet in the Solar System has its own bizarre personality, completely unlike any other. The Sun itself is disturbingly variable. And our stellar neighborhood is literally awash in effluvia from recent supernova explosions. Today, we must confront compelling evidence that most of the matter in the Universe is strangely invisible – what we call dark matter – and that the overall fabric of spacetime could be ruled by some ethereal agent which we have dubbed dark energy. Given these dizzying prospects, theoretical cosmologists, observational astrophysicists, software developers, taxi drivers, caregivers, and theologians have become our soulmates in speculating upon the cosmos that enfolds us all.

A major goal of this guide is to provide readers with a greater sense of place in the cosmos. Beginning in the first part with naked-eye views of the sky, and delving more deeply in the second, we will bear witness to the exquisite hierarchy of planets in our Solar System, the Sun as a star among myriad other stars in our Milky Way Galaxy, the Milky Way as a major player in the Local Group of galaxies, and the Local Group as a minor mote of structure in the vast galaxian firmament. Along the way, we will encounter mysterious forms of matter and energy – most of which continue to elude the most concerted scientific probing. In so doing, we will get a taste for the contentious yet progressive process of scientific inquiry.

Once the cosmic stage has been set and all the wondrous players described, the book's third part introduces us to the grand story of being and becoming. Here, readers can vicariously experience the amazing transformations that have ensued since the hot big bang, some 14 billion years ago. From the birth of galaxies out of the torrid chaos, to the formation of stars and planetary systems within these galaxies, to the emergence of life on one particularly moist planet, these transformations delineate our cosmic heritage, connecting us with all that has ever existed.

Again, scientific challenges and conundrums are explored, as this story is far from complete.

In the last chapter, readers are invited to speculate on our cosmic destiny and legacy. What will await our species – or those species which will supplant us here on this precious planet? What are the chances of our interacting with another sentient species elsewhere? And how should we comport ourselves as emerging citizens of the Milky Way?

I

OUR PLACE IN SPACE

1

First impressions

The most beautiful thing we can experience is the mysterious. It is the source of all true art and science.

Albert Einstein, *What I Believe*

Beginning with our first indelible marks on stone, we humans have incessantly expressed a keen fascination with the sky. Indigenous Australians continue to tend rock paintings with celestial themes that are thought to go back at least 15,000 years. Sometime around 2500 B.C.E., neolithic groups in England completed laying the famous stones that comprise the megalithic monument Stonehenge. The arrangement of these giant columns and lintels, along with the circle of fifty-six holes within, have suggested to some archeoastronomers that these ancient cultures used Stonehenge to mark the winter and summer solstices and to predict eclipses of the Sun.

The ancient Egyptians were especially diligent at recording on stone, wood, and papyrus what was important to them. The Sun god Ra figured prominently in many of their representations. Beginning around 3000 B.C.E., the development of hieroglyphic writing enabled the chosen scribes to elaborate upon the story of Ra and his role as arbiter of daytime life on Earth. Both the day and night skies were the province of the sky goddess Nut. Arching over the Earth and its god Geb, Nut served as the

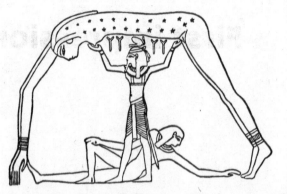

Figure 1.1 The Egyptian sky goddess Nut, vaulted over the Earth god Geb and atmospheric god Shu, c.2000 B.C.E. (Adapted from multiple sources, with reference to *The Great Goddesses of Egypt* by B. S. Lesko, University of Oklahoma Press [1999].)

corporeal host of the Sun, Moon, planets, and stars – each celestial wonder passing through her body across the vault of heaven (see figure 1.1).

By getting away from streetlights and other sources of artificial illumination, we can still bear witness to the same night sky that the neolithic Britons, ancient Egyptians, Mesopotamians, Chinese, Meso-Americans and other cultures experienced. What we can see consists of stars upon stars, some of which demarcate ragged patterns while others appear to be in tight groupings. On clear, moonless nights unsullied by light pollution, approximately 4,000 stars are available for viewing without optical aid at any one time.

In many ways the "starry dome" which we have inherited can be likened to a melting pot made up of many diverse cultural influences. For example, the names of the bright stars Sirius and Vega are of Greek origin, while Capella and Spica hark back to Roman times, and Arcturus, Deneb and Betelgeuse are anglicized versions of Arabic names. The commonly recognized constellation patterns are mostly of Greek origin but with Latin names and containing stars of varied Greek, Roman, and Arab appellations. Orion the Hunter provides a good case in point: identified by the Greeks, it contains many Arabic-named stars, including ochre-colored Betelgeuse (originally *Bet al Jauza* – the armpit of Jauza [Orion]), the three blue-white belt stars of Alnilam, Alnitak, and Mintaka, and the brilliant bluish star Rigel (the hunter's left foot).

There are historical reasons for these etymological mash-ups, of course, beginning with the first recordings of constellations by the Mesopotamians around 2000 B.C.E. The Greek philosopher Eudoxus noted forty-eight constellations in 350 B.C.E. or there-abouts, and Hipparchus carefully charted them, along with the relative brightnesses of the stars, around 129 B.C.E. The Egyptian astronomer Claudius Ptolemy (90–168 C.E.), a Roman citizen who wrote in Greek, built upon this system in 128 C.E. with a detailed catalogue of 1,022 stars which can be found in his famous treatise, the *Almagest*. Astronomers from the Arab world and central Asia – among them Ahmad al-Farghani and Abu Rayhan al-Biruni – added to the stew between 1000 and 1600 by correcting Ptolemy's errors, noting star colors and their changes, delineating borders around the constellations, and describing for the first time the neb-ular objects that could be observed with the naked eye. We now recognize those fuzzy objects to be giant clouds of gas and dust.

MEDIEVAL OBSERVATORIES IN THE MIDDLE EAST

The pre-Copernican astronomers of the Middle East and central Asia made use of major pre-telescopic observatories built during the thirteenth century in Maragha (or Maragheh) in modern-day Iran and about a century and a half later in Samarkand in modern-day Uzbekistan. One major outcome was the compilation of the most comprehensive catalogue of the heavens in the time after Ptolemy and before the European Renaissance, wherein the positions of 992 stars were recorded.

The modern Western system of constellations took form in 1603 with Johann Bayer's *Uranometria* (Measuring the Heavens), which charted for the first time both the northern and southern sky, and in 1763 with the French astronomer Nicholas Louis de la Caille's *Star Catalogue of the Southern Sky*, in which he added such non-mythic constellations as Fornax the Furnace, Antlia the Air Pump, Horologium the Pendulum Clock, and Microscopium the Microscope. Considerable confusion ensued over the extents (and in some cases overlaps) of these various constellations until 1922, when the International Astronomical Union stepped in and ordained eighty-eight "official" constellations with contiguous borders between them (see figure 1.2).

The Western system was not the only representation of the heavens to be devised, however. A completely independent system of constellations and star names had been established by the Chinese over the same span of millennia. Although this system was never adopted by the Western world, its accurate portrayal of stellar positions has been used by astronomers and archeoastronomers worldwide to pinpoint the locations of past supernova explosions, comets, and other transient celestial events. One of my favorite depictions of the night sky comes from Native American cultures over the past millennium. A comparison of their constellations with those of the Western system reveals several striking similarities and some amusing differences. For example, their version of the star pattern making up Ursa Major (the Big Bear) is known as "Three Hunters and a Bear." The Western "dog stars" of Sirius and Antares are similarly designated by Cherokee natives. Then there are the Pleiades and Hyades star clusters that reside at discrete distances from one another in the constellation of Taurus the Bull. To the Western Mono tribes, they were respectively known as "The Six Wives Who Ate Onions" and – at a safe distance – "Their Husbands."

Representations of the day and night sky continue to develop as new observing technologies are brought to bear and new celestial phenomena are discovered. So too, models of cosmic

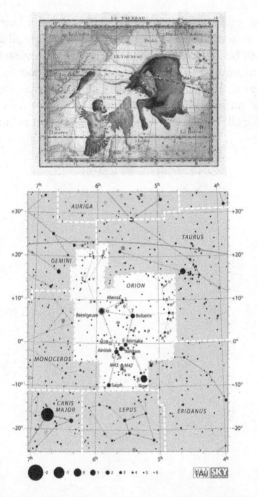

Figure 1.2 Comparison of a figurative star chart in John Flamsteed's 1776 *Atlas Celeste*, showing the constellations of Orion the Hunter and Taurus the Bull, with a more modern rendering of the same region of the sky, in which the constellation boundaries are delineated. (*Top:* courtesy of Wikimedia Commons. *Bottom:* courtesy of constellation-guide.com and *Sky & Telescope*.)

structuring continue to evolve. We will survey the known hierarchical makeup of the Universe in chapter 3. How all this came about will be deferred to part 3, where we will review the long, strange trip from the hot big bang to the emergence of galaxies, stars, planets, and life. But first, let's consider what we can see from the surface of our home planet – and how we can make sense of it all. This endeavor is commonly known as "naked-eye astronomy."

2

The day and night sky

The most incomprehensible thing about the universe is that it is at all comprehensible.

Albert Einstein, *The World as I See It*

If we could blithely fly away from Earth and into interplanetary space, we could all share the same cosmic vistas. Above us and below us, the celestial sphere would enfold us in a panoply of starlight. That is not likely to occur anytime soon, alas, and so we must contend with the fact that each of us lives somewhere on the spherical surface of a planet that is spinning about its axis every 23.93 hours and revolving around the Sun every 365.26 days.

Earthbound viewing

By being affixed to the Earth's surface, each of us can see only half of the celestial sphere at any one time, the Earth itself presenting a horizon below which nothing can be detected. Were the Earth flat rather than spherical, our visible horizon would extend all the way to the Earth's outer edge. A ship receding from us would always be visible, provided we had strong enough binoculars to resolve the ever-diminishing vessel. Instead, we

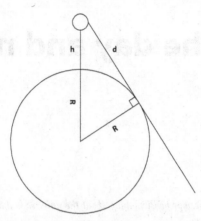

Figure 2.1 One's view of Earth's surface depends on one's height (h) above the surface along with the radius of Earth (R), as shown in this schematic. The greater the height, the farther away is the limiting horizon (d).

know that our views of receding ships have a limiting distance, and that we cannot see England from the coast of New England (see figure 2.1).

Even residents on the International Space Station are unable to see anything close to the whole (half) Earth at once. Orbiting above our heads at an average altitude of 340 kilometers (211 miles), the ISS astronauts see a curved horizon that is 2,110 kilometers away. That's about a third of the Earth's radius and a twentieth of the Earth's circumference. To do significantly better, it is necessary to move much farther away. The view of Earth from the Moon, for example, comes close to providing a completely hemispherical vista. That is why the pictures of Earth taken by the Apollo astronauts had such a transformative effect on society. For the first time in human history, the Earth could be seen in its entirety – as a precious blue orb in the blackness of space.

Why your terrestrial latitude matters

Besides limiting our terrestrial views, the Earth's curvature also causes our respective views of the day and night sky to differ. Depending on our particular latitude north or south of the equator, we will see a different unobstructed half of the sky. That is why northern observers can enjoy naked-eye and binocular views of the Pleiades and Hyades in Taurus, the Double Cluster in Perseus, and the Great Nebula in Andromeda, but miss out on similarly delightful views of the Southern Cross, Carina Nebula, and Magellanic Clouds. Those intrepid folks who venture to the North Pole would see the constellation of Ursa Minor and the pole star Polaris directly overhead. All of the other stars – down to the celestial equator – would be seen to encircle the pole star in a counterclockwise direction every 23.93 hours (see figure 2.2), while all objects with southern celestial latitudes would be forever blocked from view.

Observers at temperate northern latitudes still see Polaris due north but no longer at the zenith. Measuring the elevation of the pole star above your northern horizon turns out to be a handy way to determine your terrestrial latitude – something worth knowing if you are ever lost without GPS. Depending on your particular latitude, you would see that some constellations and the stars in them remain circumpolar, such that they never encounter the horizon. For other stellar groupings, at lower celestial latitudes the horizon definitely gets in the way. This is certainly true for constellations located along the celestial equator. A good example is the equatorial constellation of Orion the Hunter. In the months of November through March, it can be seen at night rising directly above the eastern horizon, climbing into the southern sky (for residents of the northern hemisphere), and setting directly below the western horizon.

Residents along the equator see pretty much every star in the sky rise above the easterly horizon, arc across the sky in parallel trajectories, and set below the westerly horizon. Here, the pole

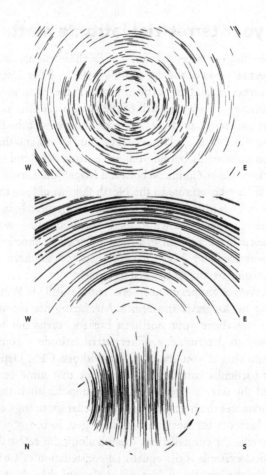

Figure 2.2 *From top to bottom:* Nighttime view of star trails from the North Pole, looking towards the zenith, compared to the view from the equator (looking north), and equivalent equatorial view looking east (or west). From the North Pole, all stars north of the celestial equator are continuously visible – never rising or setting. At the equator, the celestial poles are seen at the horizon, and there are no continuously visible circumpolar stars; instead, all stars are seen to rise and set with respect to the horizon. (Adapted from multiple sources.)

star would be barely visible, kissing the northern horizon, and Orion at its highest point would be found directly overhead. The situation for observers in the southern hemisphere is pretty much the same as that in the north but reversed. Antarctic residents would see the entire southern half of the sky but none of the northern half. Although there is (currently) no star to mark the south celestial pole, all of the stars in the sky would be seen to encircle this spot in a clockwise direction. Observers at temperate southern latitudes would see this encircling pattern displaced from the zenith towards the southern horizon, with Orion at its highest point appearing "upside down" towards the north.

HOW DO WE KNOW THE EARTH IS SPINNING?

Most people are familiar with the Sun, Moon, planets, and stars moving daily across the sky in a general east–west direction. If you ask them why this occurs, you will likely get at least a few responses referring to Earth's daily spinning about its axis. However, if you ask them how we know that Earth is spinning rather than the Sun, Moon, planets, and stars revolving daily around us, they may have difficulty coming up with a well-reasoned response. That is not at all surprising, as the evidence for Earth's spinning is not so easy to find. Indeed, most learned men prior to Galileo's time objected to the notion, as they thought a spinning Earth would cause everything that wasn't tied down to be flung off the planet. Here are a few "smoking guns" that bolster the spinning Earth hypothesis and one popular "red herring":

- **Views of Earth from the Moon.** Shortly after astronauts first landed on the Moon on July 20, 1969, they took pictures of the Earth that distinctly showed it spinning from hour to hour. But for centuries, we did not enjoy these sorts of off-planet views.
- **Foucault's pendulum.** In 1851, the French physicist Léon Foucault invented his eponymous pendulum, whereby a ponderous mass is suspended from a thin and very long line. Once set in its cyclic motion, the pendulum provides an inertial system that operates independently of Earth and its various

motions. (Gyroscopes operate in the same inertially independ-ent way.) As Earth spins beneath the pendulum, the oscillating mass is observed to sweep back and forth in Earth-based direc-tions that slowly turn with the day. Because a pendulum – if left to itself – cannot change the direction of its oscillations, it must be the Earth that is spinning.

- **Atmospheric circulation patterns.** Because Earth spins as a solid body, the rotational speed of Earth's surface decreases from a maximum of nearly twice that of a jet airplane near the equator to less than jet speed at northern and southern latitudes exceeding 60 degrees. Any air currents moving from the equator towards the poles will share the equator's high rotational velocities. As the currents pass over the slower-moving territory at higher latitudes, they will overtake the surface's eastward motion. Observers note this motion as an eastward-curving flow over Earth's surface. Flows from the poles towards the equator experience westward curving for the same reason. Combine these two curving flows and you get an atmospheric system that circulates clockwise in the northern hemisphere and counterclockwise in the south. This so-called Coriolis effect handily explains the circulation of high-pressure atmospheric systems. As air from high-pressure systems converges into regions of low pressure, its combined deflections due to Earth's spin produce a net counterclockwise circulation in the northern hemisphere and a clockwise circula-tion in the south. Most energetic storm systems are associated with these circulating zones of low atmospheric pressure.

- **The aberration of starlight.** Imagine yourself in a rainstorm prudently armed with an umbrella. There is no wind, so the rain is falling straight down. While you are standing still, you place the umbrella directly over your head, thus protecting your entire body from the downpour. But if you start walking at a good clip, you will find that the rain appears to fall at an angle from the vertical in the direction of your motion. To compensate, you will tilt the umbrella in the forward direc-tion. The same effect occurs with starlight falling on the spin-ning Earth. Stars that should be located directly overhead are observed to be shifted slightly eastward – in the direction of the Earth's overall spinning. This "aberration of starlight" amounts to only 1/11,000 of a degree. Earth's more rapid orbital motion around the Sun produces a much greater aber-ration, amounting to 1/100 of a degree. Neither of these shifts is major, but they are enough to verify the Earth's spinning and orbiting motions – and to require compensating adjustments whenever astronomers use research-grade optical telescopes to point precisely at desired points in the sky.

- **Airplane speeds.** This turns out to be a red herring. It is true that at some latitudes, the speeds of airplanes traveling eastward are greater than those traveling westward. Therefore, it would seem that the Earth's eastward spinning is enhancing an airplane's eastward velocity, while diminishing another airplane's westward velocity. However, these velocity differences can be fully explained by the eastward-directed winds that are helping to propel one airplane eastward and hindering the other airplane's progress westward. Indeed, because each airplane takes off from the spinning Earth, it shares that motion and so its airborne velocity *relative to* the surface of the spinning Earth is unaffected by the spinning.

Celestial longitudes and latitudes

Because our Earthbound views of the day and night sky are so dependent on our terrestrial location and time of day, astronomers have devised a mapping coordinate system that takes all this into account. The equatorial coordinate system is essentially a projection of Earth's grid of longitudes and latitudes onto the celestial sphere (see figure 2.3). To see how this works, consider the celestial equator. This great circle in the sky is the projection of Earth's equator outward into space. Celestial latitudes north or south of the celestial equator are measured in degrees, arcminutes, and arcseconds – where 1 degree is made up of 60 arcminutes, and where 1 arcminute spans 60 arcseconds. This is usually abbreviated as (°, ′, ″). The celestial latitude of an astronomical object is commonly called the declination – from the Latin word for "angle of the heavens." The constellation of Orion straddles the celestial equator, and so it has a declination close to 0°. By contrast, Polaris is currently located near the north celestial pole (the projection of Earth's spin axis onto the sky) and so it has a declination very close to +90°. (The Earth's magnetic axis

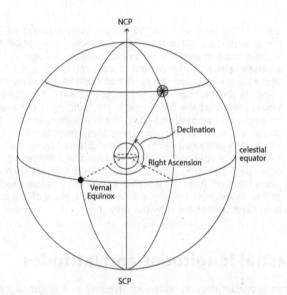

Figure 2.3 The equatorial coordinate system on Earth forms the basis for the most commonly used celestial coordinate system. Here, the inner sphere represents Earth and the outer sphere represents the celestial sphere. The celestial latitude (declination) is measured with respect to the celestial equator – with northerly declinations being positive, and southerly declinations being negative. The celestial longitude (right ascension) is measured eastward of the longitude line intersecting the vernal equinox.

continuously strays from its spin axis, with the north magnetic pole currently located at a latitude of about +86°.)

Like the lines of constant celestial latitude, lines of constant celestial longitude correspond to projections of Earth's longitudinal grid lines onto the celestial sphere. Each of these longitude lines delineates a great circle that passes through both the north and south celestial poles. Because of the Earth's daily spinning, however, it is necessary to refer the celestial longitude system to a specific location on the celestial sphere (rather than on Earth). That location corresponds to the vernal equinox – where the

Sun crosses the celestial equator every March, as it "marches" into the northern sky. Celestial longitudes are measured eastward of the vernal equinox in units of hours, minutes, and seconds – with twenty-four hours making a full circle.

This arcane use of temporal units results from the fact that Earth's daily spinning displaces the lines of constant Earth-based longitude around the sky every twenty-four hours. To determine where in the sky a particular object will appear at any particular time, it is necessary to know both the object's celestial longitude and the time of day. Keeping the celestial longitude in units of time makes the calculations more straightforward. The common astronomical term for celestial longitude is right ascension (abbreviated RA). The hours, minutes, and seconds of time (and corresponding angle) are usually abbreviated as (h, m, s). A celestial body's RA and declination (abbreviated Dec) completely describe its location on the celestial sphere. For example, the Orion Nebula in the sword of Orion the Hunter has an RA of 5h 35m 17.3s and a Dec of −5° 23′ 28″.

Keeping up with Earth's wobble

The problem with the Earth-based positions of astronomical objects is that they change slowly with time. This effect was first noted by the Greek astronomer Hipparchus around 140 B.C.E. By comparing his observations with prior astronomical records going back more than 150 years, he found that the longitudes of some bright stars along the zodiac had changed by the equivalent of 2 degrees. This amounts to an angular shifting of about 50 arcseconds per year. Given enough time, the shifting positions can really add up. For example, the location of the vernal equinox itself has shifted along the celestial equator by 28 degrees over the past 2,000 years. Around the time of the dynastic Egyptians, 5,000 years ago, the Sun crossed the celestial equator in the constellation of Aries

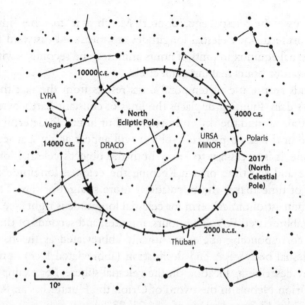

Figure 2.4 The circle in the sky around which the direction of Earth's spin axis migrates. Currently the spin axis is pointing towards Polaris in Ursa Minor. But 4,700 years ago, it was pointed toward Thuban in Draco, and in another 12,000 years, it will be pointed toward Vega in Lyra. (Adapted from Roen Kelly, *Astronomy Magazine*.)

the Ram. Today the equatorial crossing is in the constellation of Pisces the Fish. This shifting has all sorts of astrological implications. For example, just because your astrological "Sun sign" is Sagittarius, that no longer means the Sun was necessarily located in Sagittarius when you were born. Indeed, all of the astrological Sun signs are off by about one zodiacal constellation compared to the currently observed zodiacal location of the Sun in any particular month.

We now understand all this shifting of positions over time as a consequence of our spinning Earth that is also wobbling like a top. This wobbling (or precession) of the Earth's spin axis occurs over a period of about 26,000 years. Currently, Earth's spin axis is pointed towards Polaris. This direction is about 23.5 degrees away from the direction in the sky towards which the axis of Earth's orbit around the Sun is pointed. Over time, the spin axis precesses around the orbital axis while maintaining a common angular displacement of 23.5 degrees between the two axial directions. As shown in figure 2.4, the precession of Earth's axis causes the north celestial pole to migrate around the sky in a circle of 23.5-degree angular radius. While the current pole points toward Polaris in the constellation of Ursa Minor, 5,000 years ago it pointed toward Thuban in Draco the dragon. This prior "pole star" may help to explain how the architects who designed the ancient pyramids managed to align these monuments with respect to true north. They were probably referring to Thuban.

But there's more. Even the seasons on Earth have shifted with the precessional migration of Earth's spin axis. Imagine it is December 21, when Earth is in the part of its orbit where the Sun is located towards the southern constellation of Sagittarius and the north pole of Earth's spin axis is pointing maximally away from the Sun (see figure 2.5). We note this time as the winter solstice in the northern hemisphere and the summer solstice in the southern hemisphere. Now go forward 13,000 years. If you let Earth be in the same part of its orbit around the Sun, then you would find its spin axis pointed a full 57 degrees away from its current direction. With the north celestial pole now in Vega, the north pole of Earth's spin axis would be pointed *towards* the Sun on December 21. Sagittarius would now be located in the northern sky. What is today the winter solstice in the northern hemisphere would become the summer solstice (and vice versa in the southern hemisphere).

On much shorter timescales, astronomers must take into account the precessional migration of Earth's spin axis whenever they calculate the positions of objects in equatorial coordinates. This is usually done with reference to catalogues of objects whose

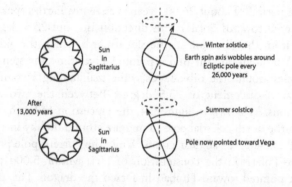

Figure 2.5 The precession of Earth's axis slowly alters Earth's relationship with the Sun. In the first case, Earth is in the part of its orbit corresponding to the December solstice, where the Sun is in the direction of Sagittarius. The northern hemisphere is curved away from the Sun and so receives sunlight at low angles – what northern occupants experience as winter. After 13,000 years, Earth's axis has precessed halfway around its cycle, so that the northern hemisphere is now pointed toward the Sun. The corresponding seasons in the northern and southern hemisphere are hence reversed, with northern summer and southern winter occurring in December.

tabulated positions are set to a particular epoch. Every fifty years, new catalogues come out. I am old enough to have referred to catalogues based on the 1950.00 epoch and, more recently, on the 2000.00 epoch. In order to determine the actual position of any object in the sky, it is necessary to "precess the coordinates" in the catalogue to the current epoch. This is usually done these days by using software at the telescope or with online calculators.

Monthly migration of the night sky

As days turn into months, a casual observer might notice that some evening stars have already set in the west by nightfall, while previously unseen stars are rising in the eastern pre-dawn sky. It is as if our entire nighttime view of the celestial sphere is slowly revolving around us in a westward direction. This intuitive interpretation has since been replaced by the modern interpretation, where the celestial sphere is kept fixed and Earth is put in motion around the Sun. The westward migration of the nighttime sky is thereby understood as the reflection of Earth's annual orbit around the Sun. As seen from Earth, the Sun gets in the way of new regions of the sky and gets out of the way of other regions (see figure 2.6).

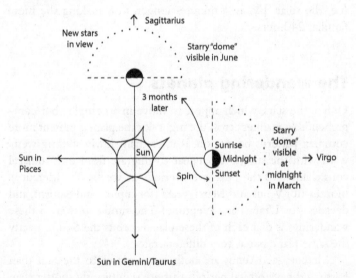

Figure 2.6 As Earth orbits around the Sun, the starry dome that is visible at night slowly migrates westward. Consider the situation where the Sun is in the direction of Pisces and the starry dome at midnight is symmetric about the direction of Virgo. After three months, the Sun is now in Gemini/Taurus, while the starry dome at midnight has shifted westward so that it is now symmetric about Sagittarius. Due to this orbital motion of Earth, new stars come into view to the east, while other stars disappear from view to the west.

The rate of shifting can be calculated based on the time it takes for the Earth to complete a full orbit. Because 360 degrees are spanned in approximately 365 days, the shifting amounts to roughly 1 degree per day. We humans are intimately tied to the Sun, and so we do not sense the shift in this way, but rather as a daily shift in the Sun-based times of star rise and star set. That time shift is the fractional shift in angle per day times a full 24 hours: $(1°/360°) \times 24$ hours $= 1/15$ hour or 4 minutes per day. This 4-minute time shift also explains why the time for the Earth to spin around once with respect to the distant stars is 23 hours and 56 minutes, whereas the spin period with respect to the Sun (i.e. the solar day) is 4 minutes longer, thus making the more familiar 24 hours.

The wandering planets

Unlike the stars, which appear to move in a complex but comprehendible manner across the night sky, the planets present more complex patterns of motion. Rather than slowly shifting westward with the stars, they appear to migrate from one zodiacal constellation to the next over periods of weeks (for Mercury), months (for Venus and Mars), years (for Jupiter and Saturn), and decades (for Uranus and Neptune). The simple answer to these wanderings is that each of these planets orbits the Sun in nearly the same plane but at very different rates.

Mercury and Venus are both much closer to the Sun than Earth and have orbital periods that are significantly shorter than that of Earth. Because they are literally closer to the Sun, we typically see these planets in the west shortly after sunset or in the east before sunrise – eliciting the misnomers "evening star" and "morning star." Mercury never appears more than 28 degrees away from the Sun and so is usually spied in the skyglow of sunrise or sunset. It cycles between the morning and evening sky and back every four months. I have seen Mercury less than a dozen times – the most memorable occurring when I was a young security guard checking the parking lot of a Lipton tea factory just before sunrise. As the factory's morning shift workers arrived, they were treated to their long-haired security guard pointing towards the eastern horizon and exclaiming, "Look, look – Mercury!" I doubt that my exhortations made them feel very secure, alas.

Venus, being almost two times farther from the Sun than is Mercury, appears to stray farther from the predawn eastern horizon or post-sunset western horizon – a full 47 degrees as seen from Earth. It migrates between these so-called greatest elongations – from the evening sky to the morning sky and back every nineteen months. That means it is possible to view Venus several hours after sunset and before sunrise, when the sky is completely dark. Venus appears so bright that it is often mistaken for an airplane – or a UFO.

Mars orbits the Sun at a distance that is 1.5 times greater than that of Earth's orbit. That means it can be seen migrating all around the great circle of zodiacal constellations. Every two years or so, the Earth overtakes Mars – like a runner on an inner lane of a circular track "lapping" a runner on an outer lane. When that happens, Mars appears to halt its usual eastward migration and reverses course, in what is known as retrograde motion. After a few more months, it halts again and resumes its eastward prograde motion (see figure 2.7).

All of the other planets exterior to Earth's orbit are observed to migrate eastward, with retrograde westward swings that last a few months, followed by resumption of their eastward motion. Again, these marvelous perambulations can be explained by Earth "lapping" these slower-moving planets every Earth year or so. Jupiter's motion across the sky is an easy one to remember. Because its orbital period is approximately twelve years, it migrates from one zodiacal constellation to the next in about a

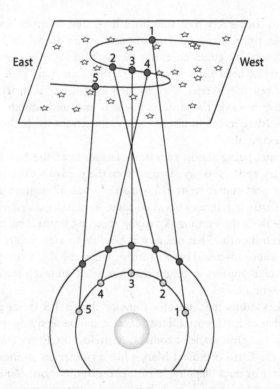

Figure 2.7 The orbits of Earth and Mars around the Sun. As Earth on its quicker inner orbit "laps" Mars, we see Mars appear to reverse course in what is known as retrograde motion (see sightlines 2–4). After a few months, Mars is observed to resume its usual eastward progression through the zodiacal constellations (see sightline 5). (Adapted from Brian Brondel, Wikimedia Commons.)

year. As I began writing this book (2015) and into the next year (2016), Jupiter was in Leo the Lion, just as it was back in 2004, 1992, and 1980 – when I first began to notice its motions around the zodiac. It has since moved through the constellation of Virgo during 2017 and on to Libra, Scorpius, etc. in subsequent years.

Jupiter's transition to retrograde motion occurs every 399 Earth days (much closer to an Earth year than that of Mars). The more distant planets take even longer to orbit the Sun, and so they pass through correspondingly smaller patches of the night sky each year, and their tiny retrograde loop-the-loops occur almost yearly (being almost a pure reflection of Earth's lapping orbital motion). We'll return to further explorations of the planets in chapter 4.

Lunar apparitions

The Moon orbits Earth every 27.3 days – its so-called sidereal period. From our own orbiting platform, we see the Moon seem to lag in its orbital motion, returning to the same place in the sky every 29.5 days. This latter synodic period, being what we observe, motivated the ancients to establish our calendrical "month" which also corresponds to one-twelfth of a full Earth year. To make it around Earth in so few days, the Moon races across the nighttime and daytime sky at a robust rate of 12 degrees per day. This motion causes the Moon to migrate through a new zodiacal constellation every 2.5 days.

But the Moon is seen to do so much more. As the Moon courses around the sky, its appearance – or apparition – changes dramatically. Consider the waxing crescent Moon, just a few days after new Moon (when it is essentially invisible). Observers with clear western horizons can clearly see this sublime lunar apparition shortly after sunset in the remnant twilight. Wait a few more days, and the Moon will have migrated eastward a full 90 degrees away from the setting Sun. You can now see the Earth-facing half of the Moon half-lit, in its so-called first-quarter phase. On this day, the Moon rises at around noon, reaches the highest point in its westward path across the sky at about 6 p.m., and sets close to midnight. Another few days' waiting will reveal the Moon in

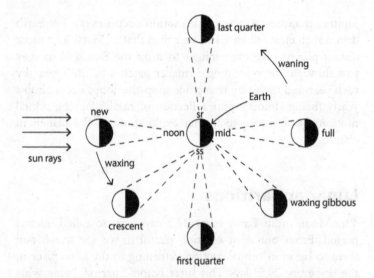

Figure 2.8 "God's Eye" view of the Moon in orbit around the Earth, with light rays from a distant Sun illuminating the two worlds. Sizes are not to scale. The Earth-based places of sunrise, noon, sunset, and midnight are noted. The dashed lines from the Earth to the Moon show what parts of the Moon are visible from Earth. For example, at full Moon, the Earth-facing side of the Moon is fully illuminated. The Moon in this phase appears highest in the sky for observers experiencing midnight. By comparison, the first-quarter Moon has only half of its Earth-facing side illuminated. It appears highest in the sky for observers experiencing sunset.

waxing gibbous phase, when more than half of the Earth–facing Moon is lit. It rises in the afternoon and sets before dawn. A full fifteen days after new Moon, if you time things just right, you can see the full Moon rising in the east as the Sun sets in the west. This exquisite apparition places you squarely betwixt the Moon and its solar illuminator – a sublime moment indeed. During full Moon, one's entire night is illuminated, as the Moon rises at sunset, reaches the highest point in its westward path at midnight, and sets at sunrise.

After full Moon, the lit portion of the Earth-facing Moon begins to decrease – producing what are known as the "waning" phases. The Moon in waning gibbous phase rises after nightfall and sets in mid-morning. The last-quarter Moon rises at midnight, reaches its highest point at sunrise, and sets at noon – an easy daytime target. Onward to the waning crescent Moon, which can be viewed in the pre-dawn hours above the eastern horizon. All of these evocative phases are the result of the theatrical lighting that the Sun provides. Being a single source of illumination, the Sun and its configuration with respect to the Moon determines how much of the Moon's Earth-facing half is lit (see figure 2.8). When the Sun and Moon are in the same part of the sky, the Moon is essentially backlit by the more distant Sun with very little of its Earth-facing side illuminated. That's when we see the crescent phases. As the Moon strays ever farther from the direction of the Sun, more of its Earth-facing side becomes illuminated. Full Moon marks that perfect moment when the Moon is located opposite the Sun, and its entire Earth-facing side is lit up like the face of a china doll.

Eclipsing delights

If the Moon orbited the Earth in exactly the same plane as Earth's orbit around the Sun, it would obscure the Sun at each new Moon – causing what is known as a solar eclipse. The Earth would also cast its shadow on the Moon at each full Moon – producing a lunar eclipse. That solar and lunar eclipses do not occur with such regularity indicates that something must be awry. In fact, the Moon orbits Earth in a plane that differs from Earth's orbital plane by about 5 degrees. That is usually enough for the new Moon to pass above or below the Sun without eclipsing it as seen from Earth. Similarly, the angular offset is enough for Earth's shadow to miss the full Moon. Eclipses do occur, however, when the new Moon or full Moon happens to be located in a part of its orbit that intersects with the Earth's orbital plane. This occurs every 1.5 years or so, producing a solar eclipse somewhere on Earth at new Moon and a lunar eclipse fifteen days later during full Moon (see figure 2.9).

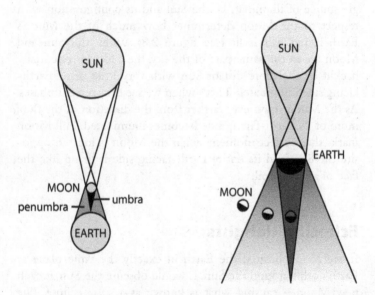

Figure 2.9 *Left:* Geometry of a total solar eclipse. People on Earth within the umbral shadow would see a total eclipse, while those within the penumbral shadow would see a partial eclipse. *Right:* Geometry of a total lunar eclipse. Here, the umbral shadow cast by Earth is wider than the Moon. (Adapted from Paul Derrick's Stargazer, www.stargazerpaul. com/s_eclips.htm.)

A total solar eclipse is one of the most spectacular cosmic scenes that can be witnessed from Earth. As the new Moon migrates precisely between the Sun and Earth, it casts a shadow that races across the Earth's surface at more than a thousand miles per hour. The umbral shadow is only a hundred miles across and so it darkens any particular spot on Earth for no more than seven minutes. During this time, you can viscerally feel the rapid plunge into darkness and then see a Sun whose heart has been blotted out. At its current distance from the Sun, the Moon perfectly eclipses the Sun. Surrounding the blackened solar disk, a diaphanous halo of luminosity delineates the solar corona that is usually overwhelmed by the Sun's brilliant surface. Beyond the corona, planets and the brightest stars can be spied in the darkened sky. And along the horizon, sunset colors trace the edge of the Moon's shadow. All too soon, the Moon migrates enough to expose a tiny glint of the solar surface – producing the dramatic "diamond ring" effect. The darkest part of the Moon's shadow has moved off, ending totality and leaving eclipse chasers to enjoy the partially eclipsed phases until the Sun once again blazes forth in full.

Compared to solar eclipses, lunar eclipses are far more sedate affairs – more suitable to enjoying over cocktails around midnight. Because the bigger Earth casts a much larger umbral shadow during a lunar eclipse than does the Moon during a solar eclipse, the entire Earth-facing Moon is darkened and for a longer time. Moreover, anybody on the nightside of Earth has an opportunity to view the Moon when in eclipse. There is no need to spend thousands of dollars to travel thousands of miles in order to get yourself situated on some special part of the Earth's surface. For this reason, you are far more likely to witness a lunar eclipse from your hometown than a solar eclipse. You will find that each lunar eclipse has its own personality. As sunlight passes through the Earth's atmosphere, some of the light scatters towards the darkened Moon. If the atmosphere is full of dust – say from a recent volcanic eruption – the light scattering will be more pronounced, and the Moon will appear a ruddy copper color. Changes in the Earth's atmospheric content will produce corresponding variations in the eclipsed Moon's brightness and hue.

To every season: turn, turn, *tilt*

I have delayed discussion of the seasons until the very end of this chapter, because in many ways the seasons are the most difficult to get right. If you ask a friend what causes the seasons, they might say that it has something to do with the tilt of the Earth's axis. So far, so good. As previously discussed, the Earth's spin axis is indeed tilted by 23.5-degrees with respect to its orbital axis. That also should mean your friend has discarded the notion that the Earth's distance from the Sun is changing enough to produce seasonal change. Earth's nearly circular orbit produces negligible changes in its distance to the Sun. And even if the distance did change by a lot, both hemispheres would be similarly affected – which is not the case. As most people know, the northern hemisphere experiences winter while the southern hemisphere experiences summer, and vice versa. But further probing often reveals that many people imagine the tilt causing one hemisphere to be significantly closer to the Sun, thus warming that hemisphere more than the other hemisphere. That's when some simple calculating is in order. Given the 23.5-degree tilt of Earth, the maximum displacement of Earth's surface towards or away from the Sun amounts to only 2,543 kilometers. Compare that with the distance of 150 *million* kilometers from the Sun, and you get an effect maxing out at 0.01695 percent. Something else must be producing the pronounced changes in climate from winter to summer and back.

Consider the situation outlined in figure 2.10. Earth is in the part of its orbit around the Sun where the northern hemisphere is tilted maximally away from the Sun and the southern

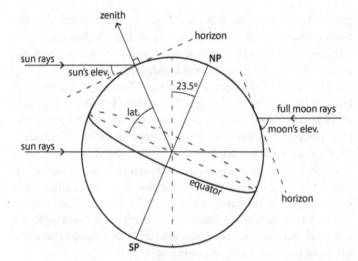

Figure 2.10 Solar irradiation of Earth on December 21, showing differences from north to south. These differences in the Sun's elevation above the local horizon – low in the sky for northern residents and high for southern – translate to a tangible difference in the Sun's heating of Earth's surface. Coincidentally, the full Moon, being situated opposite to the Sun, appears highest in the sky for northern residents, which is why wintertime midnight walks over snow-covered fields during the full Moon can be especially beautiful.

hemisphere is tilted maximally toward the Sun. Now look at the Sun's rays as they strike the surface of Earth. In the northern hemisphere, the rays strike at low angles with respect to the local horizon. A resident of the northern hemisphere would see the Sun low in the southern sky. If we liken the rays to beams of light, each beam will spread over a large area of Earth and so its luminous powering of Earth will be similarly diluted. This wan lighting of Earth's surface produces winter conditions that deepen at increasing northern latitudes. By contrast, the solar rays are simultaneously striking the southern hemisphere at high angles with respect to the local horizon. A southern resident would see the Sun correspondingly higher in the sky. At these higher angles, the luminous powering of Earth's surface would be far more concentrated – producing summertime conditions.

The angle of the Sun as it appears above the local horizon ends up being a function of one's local latitude and the degree of tilt of Earth's axis towards or away from the Sun. During the summer solstice, the noontime elevation amounts to 90° – latitude + 23.5°, and for the winter solstice it is 90° – latitude – 23.5°. For example, my home in Rockport, Massachusetts, has a latitude of +42.5°. During the summer solstice, the Sun at its highest point in the noontime sky has an elevation of 71 degrees. The resulting irradiation is enough to keep me in shorts and T-shirts. Wait a half-year and the noontime Sun rises only 24 degrees above the southern horizon. Time to bundle up!

At the spring and fall equinoxes, Earth has reached a place in its orbit around the Sun where its spin axis is tilted neither toward nor away from the Sun. With the tilting out of consideration, the elevation of the Sun ends up being purely a function of one's latitude, such that, for both the spring and fall equinoxes, elevation = 90° – latitude. In Rockport, I would see the noontime Sun at an intermediate elevation of 47.5 degrees, and my environment would receive enough solar heating to keep me comfortable in a light sweater or jacket.

Coda

In this chapter, we have considered our particular situation upon the spherical surface of a spinning and orbiting home planet while trying to make sense of our ever-changing day and night sky. Much of what we see of stars, planets, the Moon, and the Sun can only be understood with a reliable three-dimensional model in our heads – no wonder it took so long for our ancestors to figure things out! A crucial aspect of this spatial reconnoitering has been, and continues to be, the fathoming of distances to the myriad diverse objects that we can observe in the firmament. Without reckoning distances, we would be left with no more than an intriguing but unknowable assortment of luminous sources in the day and night sky. That is why job number one in astronomy is to determine ever better and farther-reaching distances to celestial objects. In the next chapter, we will survey the amazing spatial hierarchy of matter that has been revealed through this historic delineation of cosmic distances. From there, we will first briefly survey the Solar System and then move out to other stellar realms, the Milky Way Galaxy, other galaxies, and our ever-expanding galaxian Universe.

3
Cosmic perspectives

This space we declare to be infinite, since neither reason, convenience, possibility, sense-perception nor nature assign to it a limit. In it are an infinity of worlds of the same kind as our own... Thus is the excellence of God magnified and the greatness of his kingdom made manifest; He is glorified not in one, but in countless suns; not in a single earth, a single world, but in a thousand thousand, I say in an infinity of worlds.
Giordano Bruno, *On the Infinite Universe and Worlds*

Any astronomical primer that is worth reading should provide some semblance of perspective on the Universe and all that it contains. In this spirit, let's work from the inside out by considering the spatial hierarchy of matter as best we know it. We can then review the historical milestones that led us to our current understanding of the Solar System, Milky Way, and larger galaxian Universe. Lastly, we can take a vicarious powers-of-ten tour of the cosmos – from the scale of a child to the largest structures that are presently known.

Our cosmic address

In today's electronically networked world, we are used to communicating with others via addresses that are both personal and institutional. To travel somewhere or to send a physical package to some recipient, however, we must resort to geographic addresses that often include street numbers, town names, state or other regional names, postal codes, and sometimes names of countries. The same sort of addressing helps us to understand our place in space and the overall layout of matter in the Universe.

In this spirit, here is my own cosmic address:

Rockport, Massachusetts
United States of America
North America
Earth – third planet from the Sun
The Earth–Moon system
Solar System
Solar Neighborhood – featuring Sirius
Gould's Belt of bright stars
The Local Bubble
Orion Spur
Milky Way Galaxy
Local Group of galaxies
Virgo Supercluster of galaxies
Virgo–Centaurus–Hydra filament of superclusters
Laniakea
The cosmic web

This sort of addressing makes explicit the nesting of matter that pervades the cosmos; like a matryoshka doll, each component comprises part of the next component. In the remainder of this chapter, we fill in the missing "spaces" and so delineate the full hierarchical structure of matter that makes up our known Universe.

Historical milestones

It has taken humankind several millennia to fully fathom the distances to planets, stars, and galaxies. The fascinating stories of these emergent understandings make for wonderful reading (see my recommendations at the end of the book) but are beyond the scope of the present Beginner's Guide. I do think, however, that this book can and should provide a cursory introduction to the "historical milestones" that have led to our present-day picture of the Universe. So, let's start with our home planet and move out methodically, so that pertinent sizes, distances, and their historical revelations are made explicit.

Planet Earth

In chapter 2, we regarded the Earth as a spherical planet in order to understand why the night sky looks different when observed from the northern and southern hemispheres. The idea of Earth as being round was not always the accepted norm, of course, but the ancients were pretty quick to catch on once they started to document their observations of the day and night sky. One of the first records comes from the Greek philosopher Anaximander in about 600 B.C.E.: he noted that the southernmost stars visible in Egypt could not be spied from Greece. Similarly, the view from Greece featured the northern constellation of Ursa Major (the Great Bear), which completed a circuit around the pole without dipping below the horizon; in Egypt, however, the familiar constellation was seen to descend below the desert sands as part of its circumpolar circuit. Anaximander's observations were enough for him to conclude that the Earth was curved, though he ended up favoring a cylindrical shape for the world, whereby the surface curved going north–south but remained flat going from east to west.

About a century later, Parmenides provided the first cogent theoretical argument for the Earth being spherical. This follower of Pythagoras proposed that a body of any other shape than a sphere would fall inwards upon itself until it reached a state of equilibrium. According to Parmenides, a sphere was the one shape that would remain naturally in equilibrium with no further adjustments. Besides foretelling the modern view of self-gravitating bodies in hydrostatic equilibrium, such that they are neither expanding nor contracting (see chapter 5), Parmenides provided the ancients with a means for explaining where the Sun, Moon, and planets went after they set in the west. Given a spherical Earth, these bodies could keep traveling around their central host in circular orbits that would bring them back to visibility on a daily basis.

Another century later, in about 400 B.C.E., Plato famously claimed that the sphere had the most perfectly symmetric shape of any body. The Earth, being the center of the Universe, would therefore have to be a sphere. Plato's philosophical argument was based on pure aesthetics but carried a lot of weight due to his powerful advocacy of the idea.

A more empirical determination was made by Aristotle in 350 B.C.E. after he had observed the partial phases of lunar eclipses. As discussed in chapter 2, these events occur when the Sun, Earth, and Moon are all in a line, such that the Sun's illumination of Earth produces a shadow that falls on the Moon. The shape of Earth's shadow on the Moon during the partial eclipse phases appears curved – as if part of a circle. Aristotle noted this, correctly inferring that he was seeing only part of the Earth's full circular shadow.

The coup de grâce in favor of a spherical Earth was formulated by Eratosthenes in about 230 B.C.E. He used a geometric approach to derive the actual size of the Earth from measurements of the noontime Sun's elevation as observed in Alexandria and more southerly Syene (present-day Aswan). He had noted that on the summer solstice in Syene, the noontime Sun cast no shadow; it was then directly overhead. On the same day, however, the noontime Sun over Alexandria made an angle with the vertical of 7.2 degrees, or 1/50 of a full circle (see figure 3.1). By assuming a spherical Earth, he could conclude that the full circumference of Earth must be fifty times the distance between

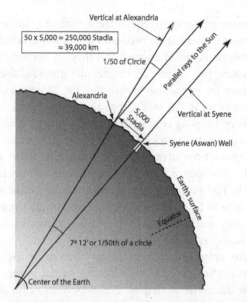

Figure 3.1 Eratosthenes's method for determining the circumference of Earth was based on observing the altitude of the noontime Sun during the summer solstice. The change in solar altitude as observed from Alexandria and Syene, along with the distance between these two cities, could be used to extrapolate the distance around the entire Earth. Here, the difference in solar altitude and corresponding distance between observing sites have been exaggerated to better visualize the salient geometry. (Adapted from the US National Oceanic and Atmospheric Administration.)

Alexandria and Syene. That amounted to 39,000–42,000 kilometers depending on his measurement, in ancient units of stadia, of the distance between Alexandria and Syene and how that distance converts to contemporary kilometers. The currently accepted value of Earth's circumference is 40,075 kilometers, so Eratosthenes appears to have been amazingly close to the mark.

By the time of Christopher Columbus, more than 1,500 years later, the idea of a spherical Earth remained, and even

Eratosthenes's derivation of Earth's circumference had managed to survive. That did not stop Columbus from adopting a circumference that was 40 percent smaller, however. Based on his reckoning, he proposed to King Ferdinand and Queen Isabella of Spain that a shorter ocean route to India was available by sailing west. King Ferdinand's expert advisors turned down this proposal, as they believed that Eratosthenes was correct and Columbus was in error. Queen Isabella overruled their decision, however, and the rest is history.

The larger Earth of Eratosthenes was finally confirmed in the eighteenth century, after teams of surveyors had slogged their way across Scandinavia, England, France, and Peru in valiant efforts to measure how long was a degree of latitude at both high and low latitudes. What they found was an average length of about 111 kilometers per degree of arc, but with variations that indicated the Earth was ever so slightly out of round. At the equator, Earth's radius measured about 21 kilometers greater than at the poles. This equatorial bulging amounts to no more than 1 part in 300. You would be hard-pressed to find a ping-pong ball so exquisitely spherical.

The Solar System

We humans have come a very long way in our engagement with the Solar System – from observing the odd motions of the planets in the night sky, to telescopically resolving the planets as individual worlds with unique identities, to remotely navigating robotic spacecraft to all of these worlds (and to many other objects) in orbit around the Sun. Key to our intellectual and practical progress was the realization that the Sun rules over all of the other bodies by virtue of its great mass and corresponding gravitation. The first person to appreciate the Sun's central role was Aristarchus of Samos – a Greek mathematician and astronomer who lived between 310 and 250 B.C.E.

Aristarchus devised a clever method whereby he could triangulate the distance to the Sun by knowing the distance to the Moon. The latter was attainable once the size of Earth had been reckoned by Eratosthenes. As Aristotle had noted, the Earth's shadow on the Moon during a lunar eclipse had a radius of curvature that was about four times larger than the Moon's radius. Therefore, the Earth must be about four times larger than the Moon (it's actually 3.7 times larger). Aristarchus took the accepted value for the size of Earth and divided it by four to derive the size of the Moon. He then compared that linear size with the Moon's angular size in the sky (about a half-degree) to geometrically deduce a distance to the Moon that was remarkably close to today's accepted value of 384,000 kilometers.

Now for the clever bit. Aristarchus had observed that the time from new Moon to first-quarter Moon was slightly less than a quarter of the Moon's perceived orbital period of 29.5 days. One way to reconcile this discrepancy was to place the Sun at a finite – and measurable – distance from the Earth–Moon system (see figure 3.2). With the Sun no longer at infinity, its rays were no longer perfectly parallel and so produced the first-quarter effect earlier in the Moon's orbit away from new Moon phase. The corresponding angle between the Sun–Earth and Earth–Moon directions could be found from the ratio of the time to first quarter to the total orbital period, multiplied by 360 degrees. Aristarchus obtained an angle of 87 degrees – about 3 degrees less than the actual angle of 89.83 degrees. That angle, when related trigonometrically to the Earth–Moon distance, can yield the Earth–Sun distance. Aristarchus preceded the advent of trigonometry by more than 100 years, but he was able to work with the geometric methods of his time to obtain a distance to the Sun that was twenty times the Moon's distance, or roughly 8 million kilometers. This was about twenty times smaller than the actual distance of 150 million kilometers, 400 times the distance to the Moon, but was enough for Aristarchus to appreciate just how far away the Sun is – and how huge it must be.

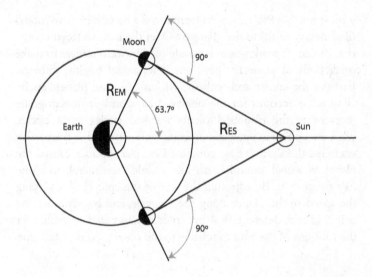

Figure 3.2 The geometric relationship that Aristarchus employed to determine the distance to the Sun based on the distance to the Moon and the position of the Moon during its first-quarter and last-quarter phases. Here, R_{EM} refers to the Earth–Moon distance, and R_{ES} refers to the Earth–Sun distance, neither of which is to scale – most important, the distance to the Sun has been purposely but inaccurately decreased so that the pertinent angles can be visualized. In actuality, the Sun is 400 times farther from Earth than is the Moon, yielding an angle of 89.83 degrees rather than the much smaller angle shown here. (Adapted from L. E. Murray.)

Aristarchus was not nearly as successful at promulgating his findings as Eratosthenes was. By the time of Nicolaus Copernicus (1473–1543), the picture of a Sun-centered Solar System had pretty much disappeared from scholarly circles. Consequently, this Polish polymath was left to build his model of the Solar System from scratch. He was motivated by the desire to improve upon the geocentric model that had begun with the Greeks and was further refined by other Middle Eastern and North African astronomers

– most notably Ptolemy, who formulated a comprehensive model of planetary orbits in the *Almagest*. Over the next thirteen centuries, Ptolemy's work was extensively used by astronomers to make predictions of planetary positions. The model begins with the Earth at the center and with the Moon, Sun, and planets encircling it. To account for the observed back-and-forth retrograde motions of the planets, Ptolemy invoked circles upon circles: what are known as epicycles orbiting deferents which themselves encircled the Earth. These combined circular motions caused the planet to wheel around Earth like a child's spirograph machine (see figure 3.3). By adjusting the size of the epicycle, by varying the speed of the planet along the epicycle, and by adjusting the origin of each deferent, Ptolemy could obtain suitable matches for the motions of the planets relative to the observations of his time.

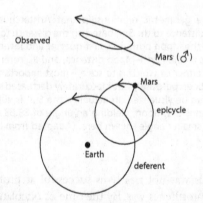

Figure 3.3   The paths of Mars, Jupiter, and Saturn across the sky undergo periodic retrograde motions, whereby each planet appears to swing westward for a few months each Earth year before reverting eastward. The geocentric model codified by Ptolemy in 150 C.E. explained the observed motion of these planets through an elaborate system of multiple circular motions. It relied on the planets moving around in circular epicycles whose centers orbited around Earth in circles known as deferents.

As naked-eye observations improved, astronomers endeav-
ored to make predictions of planetary positions in the sky over
longer periods of time. These predictions invariably failed by up
to several degrees per year. That is where Nicolaus Copernicus
stepped in, proposing his new model which he hoped would
improve upon Ptolemy's geocentric system. Perhaps building on
the work of the Persian astronomer Nasir al-Din al-Tusi (1201–
1274), Copernicus contrived a Sun-centered system of planets.
His heliocentric model was far simpler than the sophisticated
epicycles of Ptolemy and his followers. Here, the Moon orbited
around Earth, but Earth and all the other known planets orbited
around the Sun in circular paths. The Copernican model could
handily explain the yearly retrograde motions of the outer planets
as a consequence of the faster Earth "lapping" those planets (see
figure 2.7). It also enabled Copernicus to calculate each orbiting
planet's relative distance from the Sun (see figure 3.4).

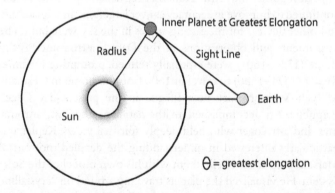

Figure 3.4 The Copernican model provided a means of calculating the
relative distances of the planets from the Sun, whereby an inner planet's
greatest observed elongation from the Sun and other favorable apparitions of the outer planets yielded key geometric relations. (Adapted from
David McClung, themcclungs.net/astronomy/concepts/plotinner.html.)

Copernicus was hesitant to promote his model. First, it displaced Earth from the center of the known Universe. This proposition was bound to provoke controversy throughout the scholarly world and – even more worrisome – displeasure from the religious authorities of his time. Second, it ended up being no more accurate in making predictions of planetary positions than the Ptolemaic system. For these reasons, Copernicus delayed publication of *De revolutionibus orbium coelestium* (On the Revolutions of the Heavenly Spheres) until 1543, just before his death. His fears of retribution were well founded. A half-century after his death, the Italian friar, philosopher, and astronomer Giordano Bruno (1548–1600) was imprisoned, tried, and ultimately convicted of heresies including the espousal of a variation of the Copernican system that allowed for an infinity of suns and solar systems (see this chapter's epigraph – in his reasoning, anything less would place restrictions on god's omnipotence). Despite appeals to Pope Clement VIII, Bruno was burned at the stake in Rome in 1600.

Throughout the early Renaissance, naked-eye observations continued to improve in accuracy as ever larger sextants, quadrants, and other devices for measuring angles in the sky were built. The instruments and observations of the Danish astronomer Tycho Brahe (1546–1601) were especially refined, prompting Johannes Kepler (1571–1630) to travel in 1600 from his home in Germany to Tycho's observatory near Prague in the present-day Czech Republic. An accomplished mathematician, physicist, astronomer, and astrologer who held deeply spiritual views, Kepler was particularly interested in understanding the detailed motions of Mars, hoping to reconcile them with his own model of the Solar System. He visualized the planets traveling around on "crystalline spheres" that were in contact with perfect Platonic solids whose sides differed in number according to particular ratios. According to Kepler's rather mystical model, the rubbing between the crystalline spheres and polyhedrons made music whose notes followed the sacred ratios which he later documented in his book

Harmonices Mundi (Harmony of the World). Through his relationship with Tycho, Kepler obtained access to the most accurate naked-eye observations of Mars then available. Tycho died prematurely of uncertain causes a year after Kepler arrived at the observatory, leaving Kepler to succeed Tycho as imperial mathematician to the court of the Holy Roman Emperor Rudolph II. Against the wishes of Tycho's family, Kepler secured Tycho's observations of planets and stars, publishing them along with his own calculations in the *Rudolphine Tables* in 1627.

Kepler's earlier publication of *Astronomia Nova* (New Astronomy) in 1609, however, is what we honor most today. Therein, he laid out two of the three laws of planetary motion that resulted from his careful analysis of the positions of Mars over time:

1. Planets orbit the Sun in elliptical paths with the Sun at one of the two focus points (see figure 3.5).
2. Planets change speed along their respective orbits, such that they sweep out equal areas in equal intervals of time. In other words, they move fastest when closest to the Sun and slowest when farthest from the Sun.

By proposing these two laws, Kepler had dispensed with the Copernican idea of perfect circular orbits of constant speed. His personal vision of crystalline spheres rubbing against perfect polyhedral solids was also sorely compromised; however, the observed motion of Mars and his own analysis of these motions required him to do these things. Kepler rose above his own misguided preferences to create one of science's greatest triumphs.

Kepler's third law of planetary motion came along after he applied similar analyses to observations of the other planets. Here, he could salvage his sacred ratios as informing the orbital periods of the planets. Published in *Harmonices Mundi* in 1627, the law

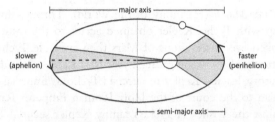

Figure 3.5 In his first two laws of planetary motion, Kepler proposed that each planet follows an elliptical trajectory with the Sun at one focus of the ellipse. The planet changes speed while sweeping out equal areas in equal times (as shown in the shaded regions, which are both the same size). Note that this figure greatly exaggerates the elongation (or "eccentricity") of the planet's orbit. The actual orbits of planets in our Solar System are nearly circular.

states that the squares of the periodic times are to each other as the cubes of the mean distances. That means the orbital periods (P) of the planets increase with increasing distance from the Sun, such that the square of the period (in years) equals the cube of the semi-major axis (a) of the elliptical orbit in astronomical units: $(P[\text{years}])^2 = (a[\text{AU}])^3$, where an astronomical unit is defined as the average distance between the Sun and Earth. Solving for the orbital period yields $P(\text{years}) = a(\text{AU})^{3/2}$.

This relationship shows that instead of swinging around the Sun in lockstep – like dirt particles on a spinning CD – the planets revolve ever slower with increasing distance from the Sun (see figure 3.6). That is why Earth periodically "laps" the outer planets of Mars, Jupiter, Saturn, etc., thus producing the observed retrograde motions of those bodies.

Simultaneous with Kepler's key findings, the Italian mathematician, physicist, and astronomer Galileo Galilei (1564–1642) was making his own inroads towards understanding the ways of the planets. Having heard about a new optical instrument developed in the Netherlands that could magnify the view of distant objects, he crafted his own "spyglasses" and started looking skyward. His

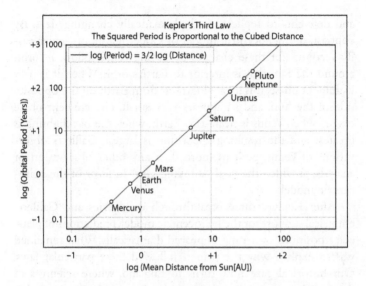

Figure 3.6 According to Kepler's third law, the orbital periods of the planets are not equal but instead increase with the 3/2 power of the mean distance from the Sun (in astronomical units [AU]). When plotted logarithmically (with powers of ten at equal intervals), this relation appears as a straight line with a slope of 3/2.

discovery of four small moons in orbit around Jupiter reminded him of a miniature Solar System and so provided a precedent for the actual Solar System being centered upon its largest member – the Sun.

Then there were his observations of Venus over its full orbit. Venus showed phases that were very similar to the phases of the Moon. This by itself could be explained by both the geocentric and heliocentric models as an effect of lighting by the Sun. However, Galileo observed that when Venus was in its crescent phase, it appeared much larger than when it was in its gibbous phase. If Venus (like the Moon) were in orbit around Earth, it would be very difficult to contrive the right sort of epicycles and deferents to replicate such dramatically changing sizes. By contrast, the heliocentric model of Copernicus and Kepler easily accounts for these changes by having Venus follow an orbit around the Sun that is interior to Earth's orbit. When it is near its fully lit phase, it is located farthest from Earth on the opposite side of the Sun, and so appears very small. The crescent phase occurs when Venus is nearest to Earth, when the backlighting is greatest and the resulting apparition is largest. Galileo's observations of Venus, perhaps more than any other observation or analysis, provided the vital "smoking gun" in favor of the heliocentric model.

After Kepler's three breakthrough discoveries and Galileo's empirical support for the heliocentric model, predictions of planetary positions over time improved dramatically. What remained was to explain why the planets followed these particular laws. Our historical gaze now turns to England, where scientists of the Enlightenment era were making great strides in many disciplines. In 1684 the mathematician and astronomer Edmond Halley (1656–1742) visited his compatriot and colleague Isaac Newton (1642–1727) at Cambridge University with a variant of this question. He asked what sort of orbit would ensue if the force binding a planet to the Sun decreased with the square of the planet's distance. Newton promptly replied that, based on calculations that he had already done, the orbit would be an ellipse. It is not clear whether Newton ever unearthed his prior calculations, but he ended up expanding upon this work to write the *Principia Mathematica Philosophiae Naturalis*, arguably one of the greatest scientific treatises of all time. In it, Newton introduced the concept of gravitational force which can act at a distance between any two masses. The force increases with the mutual magnitude of the masses and decreases with the square of the separation according to

$$F = (G \times M_1 \times M_2)/r^2$$

where r is the separation, M_1 and M_2 are the interacting masses, G is a constant of proportionality (now known as the constant of universal gravitation), and F is the resulting force. Like the situation with sound and light intensities, the gravitational force follows an "inverse square" law, such that it falls off as the square of the distance. This behavior can be understood as a consequence of space having three dimensions. More puzzling is that the gravitational force is somehow able to act remotely without any need for contact between the masses. Through its exertion, the masses are impelled to accelerate according to another of Newton's famous laws, the iconic second law of motion, $F = M \times a$, or, solving for the acceleration, $a = F / M$, where in this case acceleration (a) is the acceleration on mass (M) that is produced by force (F). By relating his gravitational force law to his second law of motion, Newton could show that each planet orbits the Sun along an elliptical path. As the planet travels closer to and farther from the Sun, it responds to the changing gravitational force by changing its velocity – accelerating and decelerating, respectively. These changes are wholly consistent with the planet maintaining its angular momentum (just as an ice skater maintains her angular momentum when she contracts her arms and spins faster). The end result is Kepler's second law of "equal areas swept out in equal times." Newton also showed that his inverse square law produced planetary orbits that vary in average speed and corresponding orbital period according to Kepler's third law. As if that were not enough, Newton showed that his relations successfully explained the motions of any massive object in the gravitational presence of another object – be it the parabolic path of a cannon ball shot from the Earth's surface or the highly elongated elliptical orbit of Edmond Halley's eponymous comet around the Sun.

Through his gravitational theory, Newton revealed the workings of the Solar System. Absolute values for the gravitational forces, distances, and masses that were involved remained elusive, however. Determination of G, the constant of universal gravitation, required the successful outcome of an experiment of exquisite delicacy. In it, a dumbbell consisting of two small lead balls separated by a rigid bar was suspended from fine wire in the presence of another fixed dumbbell consisting of two much heavier balls. The attraction and consequent swiveling of the light dumbbell toward the heavy fixed dumbbell was gauged through the torque on the twisted wire, thereby providing a measure of the gravitational force between the dumbbells. With the masses, separations, and forces known, the gravitational constant could be determined via Newton's law of universal gravitation. This was finally accomplished in the laboratory of Henry Cavendish in 1798. The modern value for G is 6.67×10^{-11} newtons (or 1 newton divided by 15 billion) of force for any two interacting kilograms of mass that are separated by a meter. Because a newton of force is equivalent to the weight of a hamburger on Earth, the gravitational constant (G) turns out to be an extremely small quantity – its effects ignorable except on astronomical scales.

With G determined, it was now possible to relate Newton's fully fleshed-out law of universal gravitation to Newton's second law of motion to make quantitative determinations of cosmic import. For example, one could use the measured acceleration of gravity at the Earth's surface (9.8 m/s^2 or 32 ft/s^2) and the Earth's known radius of 6,378 km in order to derive the Earth's mass. The resulting value of 5.97×10^{24} kilograms (5.97 trillion *trillion* kilograms) may seem colossal but it's almost negligible compared to the mass of the Sun. In order to derive that, astronomers still needed to find out the distance to the Sun – the infamous astronomical unit (AU). Instead of improving upon the long-forgotten Moon-based technique of Aristarchus, a succession of astronomers monitored the occasional transits of Venus in front of the Sun. By observing the transits from different latitudes on Earth, they could exploit geometric relations involving the respective distances between the Sun, Venus, and Earth (see figure 3.7). From

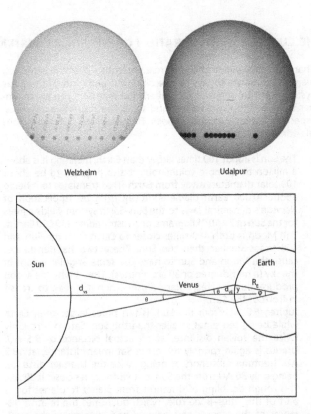

Figure 3.7 *Top:* Telescopic observations of Venus as it transited in front of the Sun on June 8, 2004. The path of the transit varies according to the latitude of the Earth-based observation. Here, observations were obtained from Welzheim, Germany (left) and Udaipur, India (right), showing subtle differences in the paths. (Courtesy of GONG/National Solar Observatory/National Science Foundation). *Bottom:* Geometric relations between the Sun, Venus, and Earth explain the variation in transiting paths across the face of the Sun according to the respective distances between the Sun, Venus, and Earth, along with the known separation between the Earth-based observations. From these relations, the corresponding distances between Earth, Venus, and the Sun can be calculated.

THE CURRENT CONFIGURATION OF THE SOLAR SYSTEM

Thanks to the valiant efforts of many scientists over the centuries, we can appreciate today the overall layout and girth of the Solar System. Here are some basic numbers that may help you to remember some of the key distances and sizes. The respective distances, especially, are for our current epoch. They may have differed billions of years ago.

- The Sun is about 100 times larger than Earth, meaning it is about a million times more voluminous. It also happens to be about 100 solar diameters away from Earth. That translates to it being about 10,000 Earth diameters away from us. Application of Newton's dynamical laws to the Sun–Earth system yields a mass for the Sun of 2×10^{30} kilograms, or an astounding 330,000 Earths.
- The Moon is both 400 times closer to Earth than the Sun and 400 times smaller than the Sun. These two happenstances cause the Moon and Sun to have the same angular extent in the sky (a half-degree or 30 arcminutes). That is why the Moon produces such perfect solar eclipses when it is seen to transit in front of the Sun.
- Jupiter, 5.2 AU from the Sun, is ten times larger than Earth while being ten times smaller than the Sun. Saturn is roughly twice the Jovian distance, at an actual distance of 9.5 AU. Uranus is again roughly twice the Saturnian distance, at 19.2 AU. Neptune falls short of being twice the Uranian distance, being only 30 AU from the Sun, but Pluto comes close at 39 AU. This rough doubling of distances from planet to planet forms part of the Titius–Bode rule, which quantifies the relative distances of the planets *and* asteroid belt in terms of the semimajor axis of their orbit in astronomical units (a), such that

$$a = 0.4 + (0.3 \times 2^n)$$

where n is the planet number moving outward, with n(Mercury) = –infinity, n(Venus) = 0, n(Earth) = 1, n(Mars) = 2, n(Asteroids) = 3, n(Jupiter) = 4, n(Saturn) = 5, n(Uranus) = 6, but replacing n(Neptune) = 7 (which fails), with n("Pluto") = 7 (which works). This arcane rule remains unexplained. Some scientists regard it as a numerical coincidence, while others say it is consistent with resonances between orbital periods that are likely to occur in an evolving planetary system such as ours.

the transits of 1761, 1769, 1874, and 1882, these intrepid astronomers determined the astronomical unit to a level of accuracy that
is within 3 percent of today's value of 150 million kilometers (93
million miles).

Solar Neighborhood – featuring Sirius

I have spent several pages introducing the key historical milestones that have led to our understanding of the Solar System as
we perceive it today. To go beyond the Solar System, astronomers
had to figure some way to gauge distances to the nearest stars.
Indeed, the concept that the stars were remote suns, and that the
Sun was our home star, critically depended on determining stellar
distances. This issue came up during the time of Galileo Galilei
as scientists grappled with the question of whether the Earth or
Sun was at the center of the known Universe. If Earth were at the
center of all things, it would be stationary, with the Sun, Moon,
planets, and stars revolving around it. As viewed from this static
perch, all of the stars in the sky would remain fixed with respect
to one another. However, if Earth were in orbital motion around
the Sun, we should observe the closest stars changing their position with respect to the more distant stars on a semi-annual basis.
This shifting of apparent stellar positions is known as geometric
parallax (see figure 3.8).

You can replicate this effect if you stick your thumb in front
of your face while looking at a distant wall with one eye open. If
you switch from one eye to the other, you will see your thumb
appear to shift in position with respect to objects on the farther wall. You will find that the shifting is greatest when your
thumb is closest to you, and that the angular displacement steadily decreases as you move your thumb farther from your face.
Similarly, the displacements made by Earth as it orbits the Sun

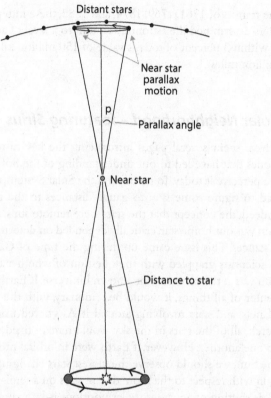

Figure 3.8 Observations of a nearby star from opposite ends of Earth's orbit around the Sun yield a shifting of that star's position with respect to more distant stars. This parallax effect can be exploited to determine geometric distances to nearby stars.

should yield shifts in the positions of nearby stars relative to those of more distant stars – the degree of shifting decreasing with the distance to the nearby star.

Tycho Brahe, Galileo, and other scientists of the Renaissance knew about the parallax effect and looked for it by tracking

the positions of bright and presumably nearby stars relative
to fainter, presumably more distant stars. Their efforts were to
no avail, however; nor were those of Robert Hooke, James
Bradley, William Herschel, and many other subsequent observers. Success was finally achieved in 1838 by the German
astronomer Friedrich Bessel, who used the finest refracting (lens-based) telescope available at the time. Sighting the
star 61 Cygni over a year, he perceived the tiniest of angular
shifts, amounting to only 0.3 arcsecond (1/12,000 of a degree).
Such a small parallax angle, along with the Earth–Sun baseline, yielded a geometric distance to 61 Cygni of 3.3 parsecs
(10.9 light years) that placed it 690,000 times farther than the
Earth–Sun distance (here, a parsec is defined as the distance
that produces a parallax shift of 1 arcsecond) In one fell swoop,
he finally confirmed that Earth moves around the Sun, *and* that
even the nearest stars are incredibly far away. Though much
dimmer in appearance, the stars are truly distant suns radiating
at comparable luminosities.

This same technique of geometric parallax has since been
employed to reckon the distances to thousands of stars. The
result has been a complete three-dimensional mapping of all
stars within 100 light years of the Earth and Solar System
(see figure 3.9). Light travels at 300,000 kilometers per second (186,000 miles/s), and so in one year it will journey 10
trillion kilometers (6 trillion miles). Our Solar Neighborhood
spans about 200 times this distance, and so there is a lot of real
estate to cover. We will get into the details in later chapters but
for now can note that most of this expanse consists of very,
very empty space. The distances between the stars are typically
50–100 million times more vast than the sizes of the stars themselves. Stretches of highway in West Texas or the outback of
Australia come to mind, but even these interminable reaches
pale compared to the interstellar emptiness that characterizes
the Solar Neighborhood.

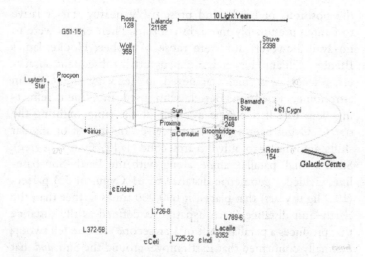

Figure 3.9 Layout of the innermost part of the Solar Neighborhood, with the Sun placed at the center. The first star to have its distance successfully measured, 61 Cygni, can be seen approximately ten light years to the right of the Sun. (Courtesy of R. Powell, *An Atlas of the Universe*.)

Much of the space in the Solar Neighborhood is also poorly illuminated, as most of the stars are dim bulbs – much dimmer in actual power output compared to our Sun. A notable exception is Sirius, the brightest-appearing star in the sky. Located 8.7 light years away in the constellation of Canis Major (the Big Dog), Sirius shines with a luminosity exceeding that of the Sun by a factor of 22. If you were trying to direct distant star travelers to your home in the Solar System, you would be well advised to tell them to aim for Sirius and then look around for a much fainter yellow star. They would see two such stars – Alpha Centauri A and the Sun. You would then have to tell them to head for the yellow star that is not accompanied by any other star, as Alpha

Centauri A is part of a triplet of stars that includes a slightly dimmer orange star (Alpha Centauri B) and a more distant and much fainter red star (Proxima Centauri).

Gould's Belt of bright stars

Beyond the Solar Neighborhood, a diadem of bright stars appears to encircle us. Indeed, some of the brightest-appearing stars in the sky can be located within this band. The British astronomer John Herschel (1792–1871) first called attention to it during his stay in South Africa, noting in 1847 a "zone of large stars which is marked out by the brilliant constellation of Orion, the bright stars of Canis Major, and almost all the more conspicuous stars of Argo [modern Puppis, Vela, and Carina] – the Cross – the Centaur, Lupus, and Scorpio." Benjamin Gould (1824–1896) was the first to trace the entirety of this feature from both northern and southern hemispheres, writing in 1874 that a "great circle or zone of bright stars seems to gird the sky, intersecting with the Milky Way at the Southern Cross, and manifest at all seasons."

The band is inclined to the Milky Way by about 20 degrees, crossing our Galaxy's gauzy midplane near the constellations of Crux in the southern hemisphere and Cygnus the Swan to the north. Nearly half of all the bright stars in the sky can be attributed to Gould's Belt. These include the red supergiant star Antares (α Scorpii) along with several other young stellar upstarts that populate the sinuous constellation of Scorpius the Scorpion; the Garnet Star (μ Cephei) – one of the largest and brightest stars in the Galaxy; the green supergiant star Mirphak (Alpha Persei), which points the way to several nearby associations of freshly minted hot stars in the constellation of Perseus the Hero; as well as the blue supergiant star Rigel and the other brilliant blue stars that adorn the familiar constellation of Orion the Hunter.

Most of the stars that make up Gould's Belt have been, until recently, too far away to have their distances gauged via the geometric parallax technique; even today, the parallax angles are too small for any measurement techniques to resolve. Instead, astronomers have carefully studied the stars that inhabit the Solar Neighborhood, where the parallax technique works, and have applied this knowledge to their observations of the more distant stars. (We will learn a bit about how stellar colors, temperatures, sizes, luminosities, and masses are determined in chapter 6.) By comparing the inferred luminosities of these stars to their apparent brightnesses, astronomers have derived distances that are good to a few tens of percent. It turns out Gould's Belt is not a perfect circle but rather an ellipse that measures roughly 2,400 by 1,500 light years. The center of the ellipse is 500 light years away from the Solar System in the direction of Taurus, roughly coincident with the Pleiades star cluster. That puts the Sun halfway between the center of the ellipse and the Belt's rim of brilliant stars. Amid the stellar luminaries in this celestial circlet, dusty clouds of atomic and molecular gas have been identified. These giant clouds are home to the next generation of stars, and so portend the near future of our Galactic neighborhood.

Gould's Belt is worthy of inclusion as part of our cosmic address for yet another reason. It is the first structure to have truly galactic dimensions. The disk of our Milky Way Galaxy measures roughly 100,000 light years across. Gould's Belt spans a respectable 2 percent of that extent. Imagine this page representing the disk of the Milky Way Galaxy. Gould's Belt would be about the size of this 0. By contrast, the Solar Neighborhood would be roughly the size of a period, and the Solar System would be a submicroscopic mote, no larger than most atoms. Extragalactic observers of the Milky Way would have to look hard with their best space telescopes, but they should be able to identify Gould's Belt.

A more recent reckoning by the GAIA satellite of distances to the stars in the vicinity of Gould's Belt has helped to constrain the distances of the associated star-forming clouds. Here, Gould's Belt melds into part of a much larger snake-like filament, which in turn comprises the local Orion spiral arm of the Milky Way Galaxy. We will have to wait for confirmation of this controversial claim, however.

The Local Bubble

Associated with the powerful stars that comprise Gould's Belt, the Local Bubble (see figure 3.10) is but one of thousands of bubbles that are thought to be percolating throughout the disk of the Milky Way Galaxy. Like its kin, the Local Bubble contains hot gas that originated from exploding massive stars. Its footprint in the Galactic disk appears to be smaller than that of Gould's Belt, and so it may be the result of only one particularly active association of stars in the Belt. Some contemporary astronomers point to a group of luminous stars in the direction of Scorpius and Centaurus as having recently hosted one or more supernova explosions that could have inflated the Local Bubble. Astronomers observing in the radio, ultraviolet, and X-ray portion of the electromagnetic spectrum have identified very tenuous and hot gas making up the bubble. This million-degree gas seems to be venting away from the disk of our Galaxy and into the so-called halo. As such, the Local Bubble could possibly be identifiable to any extragalactic observer who happens to have a clear edge-on view of our Galactic disk.

Orion Spur

Our Solar System is 4.6 billion years old, whereas Gould's Belt of luminous stars and the Local Bubble of hot gas are far more recent phenomena, having ages of only a few tens of millions of

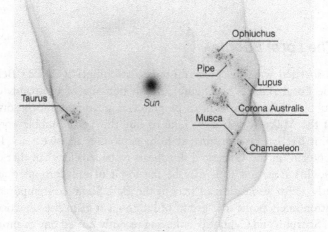

Figure 3.10   First detected at radio wavelengths in the 1970s, the Local Bubble contains hot gas that is lofting away from the disk of our Galaxy and into the Galactic halo. This artist's rendering shows young stars of our Galactic neighborhood forming on the bubble's surface. (Courtesy of CfA, Leah Hustak [STScI]/NASA.)

years. We find ourselves drifting through these newcomers like grandparents touring a nursery ward. Throughout the disk of our Galaxy, nebular matter is aggregating to create similar scenes of star formation and subsequent gaseous eruptions. Much of this seminal material is structured into vast spiral arms. We appear to be betwixt two of the major spiral arms, in what is called the Orion Spur (see figure 3.11). Astronomers first traced the Spur in the 1960s by virtue of the radio emission that its clouds of cold atomic hydrogen radiate. Further support for the feature has come from other tracers of recent star formation, including hot blue stars and the roseate nebulae which the hottest of these stars energize. All told, the Orion Spur is thought to extend for 10,000 by 3,500 light years. I must hasten caution, however, as it is notoriously difficult to surmise distances to the giant clouds of radiating gas and dust that lumber around the disk of our Galaxy. We will see shortly that certain stars have standardized luminosities that can be compared with their observed apparent brightnesses for the purposes of determining their distances and structuring, but this is not the case with the gas clouds.

Milky Way Galaxy

Astronomers have had better luck estimating the distance of our Galactic locale with respect to the Milky Way's center. The Solar System's offset location was first surmised in the 1920s by the Harvard astronomer Harlow Shapley (1885–1972) after he had determined distances to globular star clusters in the Galactic halo. Shapley based his distances on the variable stars that he could identify in the clusters. The RR Lyrae variables had a consistent average luminosity which he could use as a "standard candle" for determining distances, while the much brighter supergiant stars known as Cepheid variables had a clear relation between their luminosities and their periods of oscillation. First discovered by Harvard astronomer Henrietta Leavitt (1868–1921), this relation enabled distances to be reckoned by simply monitoring a Cepheid's variation in light output (which we'll come to shortly). From these stellar fathomings, Shapley found that the clusters were themselves "clustered" in the direction of Sagittarius the

Figure 3.11 Schematic representation of our Milky Way Galaxy based on observations at optical, infrared, and radio wavelengths. This mapping places the Sun within the Orion Spur between the Perseus and Scutum–Centaurus spiral arms. Other renderings yield a four-armed spiral pattern. (Courtesy of R. Hurt, Spitzer Science Center, Caltech/JPL, NASA.)

Archer. He correctly inferred that the nexus of the clusters' distribution traced the true Galactic center, but calculated a distance that was more than two times larger than what we accept today.

Following World War II, radio astronomers had the technological capability to improve upon Shapley's determination. By monitoring the line-of-sight motions of hydrogen gas clouds, they found a point of symmetry well within Sagittarius. On one side of this point, gas clouds were observed to be approaching us, while on the opposite side, the clouds were receding from us. Although the Galactic center is obscured from optical view behind multiple clouds of gas and dust, it could be readily observed at radio wavelengths. Detailed examination of the radio-emitting gas in orbit around the point of symmetry yielded a distance of about 28,000 light years. Distances to stars making up the surrounding Galactic bulge have produced similar values. Recently, radio astronomers have been promoting a smaller distance to the Galactic center (our Galactocentric radius) that is closer to 26,000 light years.

With this key distance pretty much known, astronomers have been able to piece together a general picture of our Milky Way Galaxy. Like other galaxies of its kind, it is a bulge–disk–halo affair that features a central bar and multiple spiral arms adorning the disk. Our most recent best guess as to the overall structuring is shown in figure 3.11. Alas, our efforts to fathom the true layout of matter in the disk remain hampered by the obscuring clouds of gas and dust that permeate it. Upcoming space missions may ameliorate this perceptual impasse, so best to stay tuned.

Not shown in figure 3.11 is the extended halo of dark matter that is thought to completely permeate and enshroud the Milky Way. Such an invisible halo has been invoked in order to explain the puzzlingly high orbital velocities of gas in the Milky Way's outer disk. Something must be gravitationally binding that speedy gas so that it does not escape from the Galaxy. What that "something" is remains completely unknown. Even more disconcerting, this invisible substance is thought to comprise more than 85 percent of the Milky Way's total mass!

Local Group of galaxies

The Milky Way Galaxy is not alone (see figure 3.12). Its irregular companions – the Large and Small Magellanic Clouds (LMC

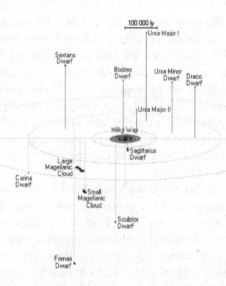

Figure 3.12　The Milky Way Galaxy is surrounded by the Large and Small Magellanic Clouds along with about ten other, much smaller, dwarf galaxies. (Courtesy of R. Powell, *An Atlas of the Universe*.)

and SMC, respectively) – have been known as naked-eye highlights of the southern sky since well before Ferdinand Magellan's voyage around the world in 1519–22. Reliable distances to the Magellanic Clouds had to wait for astronomers to develop giant reflecting telescopes and photographic technologies capable of imaging individual stars in the Clouds. From long-exposure photographs of the Small Magellanic Cloud taken over many days, Henrietta Leavitt identified in 1908 several particularly bright stars therein, known as Cepheid variable stars. Varying in brightness over days to weeks, these supergiant stars revealed to Leavitt a clear relationship between each star's period of fluctuation and its absolute luminosity. By comparing these stars to Cepheid variable stars within the Milky Way itself, Leavitt and the Swedish astronomer Ejnar Hertzsprung obtained a distance to the Small Magellanic Cloud that was truly extragalactic. Today's distances of 200,000 light years to the SMC and 160,000 light years to the LMC place these galaxies in the outermost fringes of our Milky Way's extensive dark matter halo.

Deep digital images from the most powerful telescopes have revealed more than ten galaxies in association with the "greater" Milky Way. Most of these galaxies are relatively insubstantial "dwarfs" of irregular or ellipsoidal shape. They are thought to represent vestiges of a primeval period when the Milky Way was condensing from a dense swarm of dwarf galaxies. Another ten or so dwarf galaxies are buzzing around the Andromeda Galaxy (M31), the next nearest giant spiral galaxy. M31 is currently located 2.5 million light years away, but is expected to merge with the Milky Way 3 to 5 billion years from now. Beyond the Milky Way and M31, several outliers complete the Local Group's dossier of galaxies. These include the Pinwheel Galaxy in Triangulum (M33), a small but impressively active star-forming spiral galaxy; and IC 10, the nearest starburst galaxy – where the rampant starbirth and stardeath activity has led to a galaxy-wide filigree of excited gaseous structures.

Virgo Supercluster of galaxies

Deep telescopic surveys of the sky have revealed galaxies upon galaxies – often in loose congregations akin to the Local Group. Some of the concentrations are richer in galaxies, qualifying those groups for the term "cluster." The Virgo Cluster, a peppering of galaxies that sweeps across the entire constellation of Virgo the Virgin, is the nearest such structure. Determining distances to the galaxies making up this cluster was problematic until fairly recently. The largest ground-based telescopes could not distinguish the Cepheid variables among the background of other stars, and so secondary tracers of luminosity and distance had to be invoked. These included the brightest star-forming nebulae and globular star clusters that appeared in long-exposure images. Beginning in the 1990s, the Hubble Space Telescope (HST) was tasked with imaging galaxies out to and including the Virgo Cluster galaxies. The HST's superior acuity was sufficient to identify Cepheid variables in these galaxies, monitoring their periodic variations in brightness, and so giving astronomers the wherewithal to determine their distances. Their latest determinations place the centroid of the Virgo Cluster 54 million light years away, but with member galaxies that stray from this centroid by as much as 7 million light years.

Containing more than 1,300 bright galaxies and an unknown number of much fainter dwarf galaxies, the Virgo Cluster manifests what can happen when galaxies evolve in close proximity with one another. The outer zone of the cluster contains several giant spiral galaxies not too different from the Milky Way. Closer to the ill-defined core, the galaxies resemble ellipsoidal star swarms rather than disk-based spirals. Astronomers attribute the mostly featureless stellar pileups to collisions between galaxies that scrambled the galaxies' stellar orbits and – in some cases – led to major mergers between them.

The Virgo Cluster is but one of more than 100 galaxy groups and clusters that populate the Virgo Supercluster. The Milky Way and its brethren Local Group of galaxies are all outlying members of the Virgo Supercluster. Most of the galaxies in this far-reaching structure are much too remote for their stars to be resolved and fathomed in terms of distances. Instead, astronomers have exploited the relentless expansion of the Universe to gauge the extent by which the light waves from these galaxies have been stretched since they were first emitted. According to the law of universal expansion first articulated by Georges Lemaître in 1927 and observationally confirmed by Californian astronomer Edwin Hubble in 1929, the degree to which a radiating galaxy's light waves are stretched (often referred as the galaxy's redshift) is directly proportional to the distance of the galaxy from Earth. By pegging the redshift–distance relationship to galaxies of well-known distance (as obtained from the fluctuation periods of their Cepheid variables), astronomers have since refined the so-called Hubble Law of universal expansion for the purpose of gauging distances to galaxies throughout the Virgo Supercluster and beyond.

What they have found is a loosely knit assemblage of galaxy groups and clusters that is roughly centered on the Virgo Cluster and that extends for about 110 million light years. The thousands of bright galaxies account for 3 trillion suns of matter, but the overwhelming dark matter could increase the total mass to the equivalent of about 1,000 trillion suns. Relative velocities among the sundry galaxian components are of the order of 500 kilometers per second. At this rate, the time for a galaxy to cross the Virgo Supercluster would be 66 billion years – five times the age of our Universe. That suggests the Virgo Supercluster has a long way to go before it responds to its own gravitation. In other words, this structure is primeval, still congealing from the hot big bang.

Virgo–Centaurus–Hydra filament of superclusters

Since the 1980s, optical astronomers have dedicated thousands of nights of telescope time to ascertaining the spectroscopic redshifts – and corresponding distances – of myriad galaxies throughout the sky. At first, the required spectrographic observations were obtained one galaxy at a time. But new technologies have emerged whereby the light from hundreds of galaxies in a cluster can be simultaneously dispersed into individual spectra and captured *in toto* on electronic array detectors. The resulting three-dimensional distributions of galaxies have yielded tantalizing yet puzzling glimpses of large-scale structure in the Universe.

Some astronomers see vast sheets of galaxies that are organized into bubble-like surfaces. Others liken the large-scale structure to a cosmic cobweb of galaxies and associated dark matter that consists of tenuous "filaments" converging towards denser "nodes" where the superclusters reside. Our Milky Way Galaxy appears to be part of the Virgo–Centaurus–Hydra filament of superclusters. Besides galaxies and the elusive dark matter, the filaments also contain gobs of hot gas whose presence has been surmised from sky surveys at ultraviolet and X-ray wavelengths. For now, we don't know what all this structuring means, but are heartened by the fact that it is consistent with the spacing of blotches that is evident in recent mappings of the cosmic microwave background – the afterglow of the hot big bang. Somehow, the distribution of galaxies on the largest scales is telling us how matter was organized on the smallest scales in the first nanoseconds following the emergence of our Universe.

Laniakea and beyond

The largest structure to which we belong is known as Laniakea – meaning "Immense Heaven" in Hawaiian. It encompasses the Virgo–Hydra–Centaurus, Fornax–Eridanus, and Pavo–Indus superclusters. A great void separates Laniakea from the next big galaxian system that is centered on the Perseus–Pisces Supercluster. Together, they span more than a billion light years. Beyond these structures, our sample of galaxies is too incomplete to infer much. We can see a few galaxies hosting brilliant quasar phenomena and gamma-ray bursts, along with a smattering of starbursting primordial galaxies, at redshifts that place them up to 10 billion light years away – but certainly not enough of these beasts to generate well-populated three-dimensional maps. New telescopic capabilities and survey methods may soon fill in this gap in our galactic census, thus enabling us to trace both the structure *and* evolution of matter throughout the observable Universe.

Cosmic zooms

For most of this chapter, we have considered the spatial hierarchy of nesting structures in the cosmos and how astronomers came to fathom it all. In some ways, it's all too much to ponder at once. That's where a bit of mathematics can help. A handy way to deal with the incredible range of sizes and distances in the Universe is to start with some readily measurable scale (such as a meter) and multiply by ten, multiply by ten again, and again, etc. Doing so will give you a "Powers of Ten" tour of the Universe that will take you to the edge of all space (and time) in only twenty-six steps.

10^0 m = 1 meter (four-year-old child, large stride of adult, boa snake, large meteorite)

10^1 m = 10 m (whale, large classroom, Hubble Space Telescope)

10^2 m = 100 m (redwood tree, algae bed, football field, Tunguska comet fragment)

10^3 m = 1,000 m = 1 kilometer (school of shrimp, university campus, volcanic eruption of Mt. Saint Helens [1 km^3])

10^4 m = 10^1 km (Boston, thickness of Earth's biosphere, Martian satellite Deimos, comet nucleus, neutron star)

10^5 m = 10^2 km (Massachusetts [N–S], most known asteroids, Uranian satellite Miranda)

10^6 m = 10^3 km (Gulf of Mexico, largest asteroid Ceres, the Moon)

10^7 m = 10^4 km (Earth, white dwarf star Sirius B)

10^8 m = 10^5 km (Earth–Moon distance [4×10^5 km], Jupiter)

10^9 m = 10^6 km (the Sun, Sirius A, and other "normal" hydrogen-burning stars, comet coma)

10^{10} m = 10^7 km (comet hydrogen halo, blue giant star Rigel)

10^{11} m = 10^8 km (comet tail, red supergiant star Betelgeuse, Earth–Sun distance [1.5×10^8 km] = 1 astronomical unit [1 AU])

10^{12} m = 10^9 km ~ 10 AU (Sun–Saturn distance: 9.2 AU)

10^{13} m = 10^{10} km ~ 10^2 AU (diameter of Pluto's orbit ~ 80 AU)

10^{14} m = 10^{11} km ~ 10^3 AU (planetary debris disk around Beta Pictoris)

10^{15} m = 10^{12} km ~ 10^4 AU (distance from the Sun to the Oort Cloud of comets: 5,000–50,000 AU)

10^{16} m = 10^{13} km ~ 10^5 AU (distance light travels in one year: 1 light year [l.y.] = 9.5×10^{15} m = 9.5×10^{12} km = 6.3×10^4 AU, distance to Proxima Centauri = 4.2 l.y. = 2.6×10^5 AU)

10^{17} m ~ 10 l.y. (distance to nearest 30 stars in Solar Neighborhood, size of molecular cloud core, size of the Orion Nebula)

10^{18} m ~ 10^2 l.y. (radius of Solar Neighborhood, distance to nearest molecular cloud [4×10^2 l.y.], size of giant molecular cloud)

10^{19} m ~ 10^3 l.y. (distance to nearest giant molecular cloud [1.5×10^3 l.y.], thickness of Milky Way's disk)

10^{20} m ~ 10^4 l.y. (width of spiral arm in Milky Way)

10^{21} m ~ 10^5 l.y. (size of Milky Way Galaxy)

10^{22} m ~ 10^6 l.y. (distance to Andromeda Galaxy [2.5×10^6 l.y.])

10^{23} m ~ 10^7 l.y. (distance to most other galaxies in the local Universe)

10^{24} m ~ 10^8 l.y. (size of Virgo Supercluster of galaxies)

10^{25} m ~ 10^9 l.y. (distance to quasars and primeval galaxies)

10^{26} m ~ 10^{10} l.y. (distance to the visible "edge" of the Universe … to epoch of the hot big bang)

Several cosmic zooms have been rendered as video animations. These include the pioneering *Powers of Ten* tour of the macroscopic and microscopic Universe by Charles and Ray Eames (see www.powersof10.com/film), the interactive *Scale of the Universe* graphic (see www.scaleoftheuniverse.com), part of a *Simpsons* TV episode, and the fanciful ending to the first *Men in Black* movie.

Now that we have journeyed out to the edge of the known Universe, let's now take deeper dives into the respective realms, beginning with our Solar System – the only stellar and planetary system that we can directly explore.

II

CONSTITUENTS
OF THE COSMOS

4

Introducing our Solar System

Outside intelligences, exploring the solar system with true impartiality, would be quite likely to enter the sun in their records thus: Star X, spectral class G0, 4 planets plus debris.

Isaac Asimov, Essay 16, "By Jove!," *View from a Height*

If this book had focused exclusively on the Solar System, and its contents were apportioned by mass, every page would be full of information on the Sun, as it contains 99.86 percent of the Solar System's total mass. The planets, asteroids, plutoids, and sundry comets that make up the rest of our Solar System would all fit on the last half-page. Instead, this book considers the Sun and its horde as but one system among a panoply of other stellar and planetary realms that make up our Milky Way Galaxy. So, let's get to know our home system a little better – for its own sake, of course, but also to have some reference for exploring what lies beyond.

Basic plan

The overall layout of our Solar System can be visualized in a variety of ways. For example, the ten–billion–to–one scale model situated on the National Mall in Washington, D.C. has the Sun about the size of a grapefruit, with Earth a poppyseed 15 meters away, Jupiter a big blueberry 78 meters away, and Neptune a peppercorn 430 meters distant. This sort of "linear" visualization helps to highlight the vast spaces between the planets relative to their sizes. Another way to visualize the Solar System is to compress the distances logarithmically, so that you can more easily cope with the full range of distances that are traced by the planets and outer comets. This sort of rendering is shown in figure 4.1.

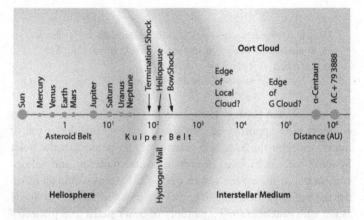

Figure 4.1 Spatial distribution of planets, asteroids, and comets on a logarithmic scale of distances – with powers of ten at equal intervals. The logarithmic scaling enables the enormous dynamic range of distances to be compressed into a single map. The rough doubling of planetary distances according to the Titius–Bode rule is manifested here as equally spaced intervals. The asteroid belt of rocky bodies is most populated in the zone between Mars and Jupiter. The Kuiper Belt refers to the many icy bodies beyond the orbit of Neptune – these include Pluto, its kindred plutoids, and much smaller comets. Beyond the Kuiper Belt, the Solar System extends halfway to the nearest star system (Alpha Centauri). The Oort Cloud of comets occupies this most remote part of the Solar System. While Saturn is impressively ten times farther from the Sun than is Earth, the Oort Cloud is yet another 5,000 to 50,000 times farther away. (Adapted from Wikimedia Commons.)

OBSERVING OUR SOLAR SYSTEM

Much of what constitutes the Solar System defies ready reconnaissance, yet there remains an awful lot that can be perceived by just looking up. Indeed, there is nothing like viewing these marvels for yourself.

Viewing the **Sun** presents a safety hazard without precautions. One of the safest ways to view the Sun is through projection. Using a cardboard box, you can build a simple pinhole camera by cutting a 2-inch (5-cm) hole out of the middle of one side and covering it with a piece of aluminum foil that has been pierced once by a pin. That hole, when pointed toward the Sun, will then project an image onto the opposite wall of the box. Another method is to view the Sun through mylar filters with reflective coatings. These can be purchased in the form of "solar shades" and as custom-fitted filters for a telescope. Finally, there are stand-alone telescopes that filter out all light but a particular red emission produced by the Sun's stormy outer layer known as the chromosphere.

Mercury and Venus appear in the sky as "morning stars" or "evening stars," meaning they are often found in the east just before sunrise or in the west just after sunset. The best time to view Mercury is when it is near its point of greatest elongation from the Sun. This amounts to no more than 28 degrees, so Mercury is always found in a significantly brightened sky. Venus has a far more forgiving maximum elongation of 47 degrees, so is possible to see in a completely darkened sky. Through a small telescope, Mercury reveals very few of its secrets, but again Venus is far more accommodating, such that one should be able to discern its particular phase.

Mars, Jupiter, and Saturn all orbit the Sun well beyond the Earth's orbit. That means they are not limited in their elongations from the Sun as seen from Earth. Indeed, the best times to view these planets are when they are completely opposite the Sun, when they appear highest in the sky at local midnight. This state of opposition occurs roughly once every 780 days for Mars, 399 days for Jupiter, and 378 days for Saturn. The angular sizes of the planets are largest during opposition, as the Earth and these planets are then all on the same side of the Sun and so are closest to one another in their respective orbits. Naked-eye views of Mars can reveal a ruddy hue, while those of Jupiter are much whiter, with Saturn yielding a tawnier tone.

Binocular views of Mars provide a better rendering of its rusty color but not much more. A properly focused refracting or reflecting telescope during times of opposition will typically show one or both of the polar ice caps along with some darkened features such as Syrtis Major. A well-aligned pair of binoculars will show Jupiter's substantial girth along with its four Galilean moons. Amateur telescopes can bring into fascinating focus the giant planet's dark belts and bright zones, along with smaller features – including colorful curlicues, bright and dark spots, and the Great Red Spot itself.

Being twice as far as Jupiter, Saturn reveals considerably less through binoculars. Its tallow color becomes more conspicuous, while its largest moon, Titan, is readily evident. Some will be able to see some asymmetries in Saturn itself, just as Galileo saw ear-like protuberances with his spyglass more than 400 years ago. It takes a decent telescope, however, to resolve these extensions into the exquisite ring system for which Saturn is justly famous.

Uranus and Neptune are both beyond the reach of the unaided eye. High-quality binoculars and a good sky chart are sufficient to spot Uranus, but a telescope is required to sight Neptune. You will see no more than shimmering dots, with Uranus appearing greenish and Neptune appearing more bluish. Considering that these major planets were completely unknown at the time of the American Revolution, you will likely cherish whatever views you get of them.

Each decade we are witness to several **comets** that come close enough to the inner Solar System to put on magnificent displays. The most recent flyby of naked-eye prominence was of Comet NEOWISE in 2020. The most spectacular cometary apparitions feature a brilliant nucleus, from which a straight plasma tail and a curved dust tail emanate. Telescopes will enable you to zoom in to the nucleus and surrounding coma, where you may see evidence of curving streamers – provided you have situated yourself well away from sources of light pollution, that is.

The terrestrial planets

The innermost planets – Mercury, Venus, the Earth–Moon system, and Mars – all have average densities that are consistent with them being made almost entirely of rock. They are also considerably smaller than the gas giants (Jupiter and Saturn) and the

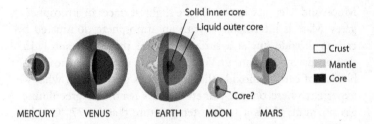

Figure 4.2 Cross-sectional depictions of the terrestrial planets and our Moon. Each planet contains a metal core (of varying size), mantle, and crust. (Adapted from NASA.)

ice giants (Uranus and Neptune) that populate the outer Solar System. Perhaps their most distinguishing feature is their close proximity to the Sun. Warmed by abundant sunlight when they were first forming, the terrestrial planets were likely prevented from accreting and holding onto the light gases of hydrogen and helium that comprise the bulk of Jupiter and Saturn. These gases would have been too volatile in the Sun's energizing presence to stay put. What remains are uniquely fascinating worlds – more disparate and puzzling than one could have ever imagined. Figure 4.2 shows the interiors of each object. Mercury contains an anomalously large metal core that occupies 75 percent of its radius and 42 percent of its volume. By contrast, the Moon sports a negligible metal core. Venus and Earth are thought to have similar substructures, with cores that occupy about 55 percent of their radii and 16 percent of their volumes. Mars is intermediate in size between Mercury and the near-twins of Venus and Earth.

The magmatic activity at the surfaces of these bodies appears to increase with their size. For example, the Moon, Mercury, and Mars have histories of prior activity but not much of anything going on in the last billion or so years. Earth and Venus, on the other hand, continue to actively transform their surfaces. The atmospheric content also appears to trend with mass. The

Moon and Mercury have only the slightest traces of atmospheric gases. Mars is intermediate, with an atmosphere dominated by carbon dioxide but at a surface pressure that is a scant 1/157 that on Earth. By contrast, Venus is buried under an atmospheric blanket of carbon dioxide that is 93 times greater than what we experience here on our planet. For this reason, its greenhouse-gas warming yields a surface temperature that is 477 °C – hot enough to melt lead and vaporize sulfur.

The gas giants

Like bullies on the block, Jupiter and Saturn throw their weight around the Solar System. Jupiter, especially, is likely responsible for having prevented the assembly of a bona fide planet in the zone currently occupied by the asteroids. Instead, most of the sub-planetary bits that once occupied this annular zone were gravitationally perturbed by Jupiter out of the Solar System or into the Sun – leaving the relatively paltry population of aster-oids that we find today. Jupiter has also had its way with com-ets – depopulating them from their birthplaces near the orbit of Neptune and redirecting them into the inner Solar System or out to the Oort Cloud of comets some 5,000 to 50,000 AU from the Sun. These giant worlds are interesting in their own right, of course, while their ring systems and coteries of satellites are noth-ing less than amazing.

Jupiter

Named after the Roman king of the gods, Jupiter contains more than three-quarters of all planetary matter in the Solar System. We know this because we can monitor the orbital motions of its moons and, with knowledge of their distance from Jupiter, can use Newton's law of universal gravitation to ascertain the gravitating mass responsible for binding these motions. The resulting mass of 318 Earths, together with a volume that would engulf 1,405 Earths, yields an average density of 1.34 grams per cubic centimeter. This is roughly one-third the mean density of rocky Earth, and very nearly the density of water in Earth's oceans (1 g/cm^3). In truth, Jupiter could be better described as a 'liquid giant' planet rather than a gas giant. Its interior structure remains fairly uncertain but is thought to contain a relatively dense core of unknown composition, surrounded by successive layers of liquid atomic hydrogen so dense that it conducts electricity like a metal, liquid molecular hydrogen, and gaseous molecular hydrogen, ammonia and water (see figure 4.3). All this spins around every 9.8 hours, thus driving intense winds in the planet's atmospheric belts, zones, and spots. Orbiting this whirling giant are at least seventy-nine moons, including the four Galilean moons – Io, Europa, Ganymede, and Callisto – each of which can be regarded as an astonishing world in itself.

Saturn

Through a decent telescope, the sight of Saturn's prominent ring system is nothing short of amazing (see figure 4.4). Many people marvel at the seeming impossibility of what they have just seen – as if somebody had suspended a model in front of the telescope. The planet itself resembles Jupiter in many ways. It shares Jupiter's Sun-like composition, with hydrogen making up the vast majority of its volume (but with helium strangely depleted). It also shares Jupiter's rapid rotation, where one Saturnian day takes only 10.6 hours. This zippy spinning has roiled Saturn's atmosphere into banded circulation systems akin to the parallel dark belts and bright zones on Jupiter. Also like Jupiter, Saturn is still radiating away the energy that was released

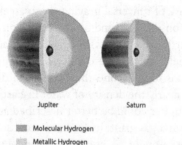

Jupiter   Saturn

Molecular Hydrogen
Metallic Hydrogen

Figure 4.3   *Top:* Schematic cross-section of the gas giants' interiors, where the different layers are based on physical models. In each case, the nature of the core is the least certain. *Bottom:* Image of Jupiter by the Cassini spacecraft, showing dark belts, brighter zones, white ovals, and the Great Red Spot. The black dot is the shadow of Jupiter's second-closest satellite, Europa. (*Top:* courtesy of NASA/Lunar and Planetary Institute; *bottom:* courtesy of NASA/JPL/University of Arizona.)

during the gravitational collapse of its birth cloud. Indeed, the two planets radiate more heat energy than they receive from the Sun.

Figure 4.4 Saturn and its icy rings present one of the most stunning sights in the Solar System. Because of the 27-degree tilt of its spin axis with respect to its orbital axis, Saturn changes in appearance as viewed from Earth over the course of one Saturnian year (29.5 Earth years). As shown in this Hubble Space Telescope composite, from 2000 to 2006 Saturn's ring system opened up from being just past edge-on to being nearly fully open with its southern hemisphere experiencing summer. (Courtesy of NASA/Hubble Heritage Team/STScI.)

Saturn's ring system is thought to be composed of various ices arranged into concentric bands of exquisite thinness. The rings could be evanescent – coming and going on timescales of 100 million years or so. Beyond the rings, more than sixty moons orbit the planet – including Titan, which is the only moon in the Solar System to host a substantial atmosphere, and Enceladus, which has liquid geysers erupting from its surface.

The ice giants

The ancients were fully aware of Mercury, Venus, Mars, Jupiter and Saturn, as all of these planets could be spotted and tracked by the unaided eye. It took the development of the telescope, however, to augment these six major planets (including Earth) with the more remote and much fainter worlds of Uranus and Neptune. Together, Uranus and Neptune represent the outermost major planets in our Solar System. Pluto was officially demoted in 2006 to minor or "dwarf" status due to its much smaller size, while searches for another major "Planet X" have (so far) come up dry.

Uranus

Uranus orbits the Sun at a distance of 19.2 AU – four times farther than Jupiter and twice as far as Saturn. Therefore, it receives 1/16 the radiation received by Jupiter and 1/4 that illuminating Saturn. At an equilibrium temperature of −208 °C, the ammonia and water in the Uranian atmosphere have completely frozen out and "snowed" down to lower levels. What remains in the atmosphere is a surfeit of molecular hydrogen and methane gas. The methane preferentially absorbs red light from the Sun while reflecting the green and blue light. This biased reflectivity causes the visible atmosphere of Uranus to appear greenish.

The planet's mass (fourteen times that of Earth) and size (four times that of Earth) yield an average density of 1.3 g/cm^3. Planetary scientists infer from these bulk properties that Uranus contains a metal–rock core, a thick intermediate layer of icy slurry, an outer layer of liquid molecular hydrogen, and a thin atmosphere (see figure 4.5). Therefore, the moniker "ice giant" appears to be fairly apt. The above conclusions are fairly recent, though, as books on the planets dating back to 1995 claimed that

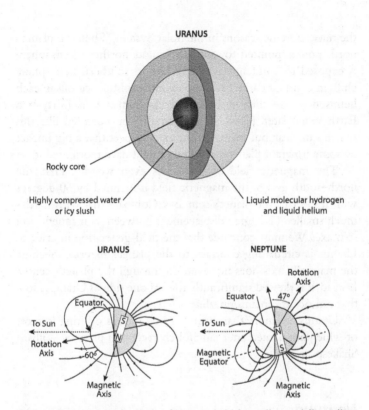

Figure 4.5 Interior schematic of Uranus (which applies equally well to Neptune) and the odd configurations of the rotational (spin) axes and magnetic fields in these two "ice giants." (*Top:* adapted from *Astronomy* by C. J. Peterson; *bottom:* adapted from NASA's Cosmos.)

the interiors of both Uranus and Neptune were dominated by liquid hydrogen.

Very little was known about Uranus until Voyager 2 whipped past it in 1986. During the spacecraft's brief rendezvous, it was confirmed that the spin axis of Uranus is tilted by 98 degrees so that it is nearly lying on its side. Such an extreme tilting leads to the most extreme seasons in the Solar System. When the planet's north pole is pointed toward the Sun, its northern hemisphere is exposed to constant daylight, while its southern hemisphere chills in constant darkness. These extreme solstice seasons in each hemisphere are then reversed a half-Uranian year (forty-two Earth years) later. How Uranus came to be upended like this remains unclear, but many astronomers suspect that a big impact by some itinerant planet-sized body might have reoriented it.

The magnetic field of Uranus is even weirder. First, the north–south axis of the magnetic field is inclined by 60 degrees with respect to the planet's spin axis. Compare that with Earth's much smaller 11-degree displacement between its magnetic and spin axes. We must conclude that the field-generating interior of Uranus is circulating contrary to the planet's exterior. Second, the magnetic axis does not even go through the planet's center. Instead, it is skewed significantly toward one side of Uranus. How that can occur remains puzzling.

Uranus has more than twenty-seven moons orbiting it, most of which are named after fanciful characters in plays by William Shakespeare.

Neptune

I fondly remember August 25, 1989, when Voyager 2 finally made it to Neptune after having traveled through interplanetary space for more than twelve years. I was then working as a lecturer in astronomy at the University of Washington in Seattle. There was no internet back then, and the television stations would dedicate very little airtime to the epochal flyby. Fortunately, the university had as part of its audio-visual services a television monitor that could receive the transmissions from NASA's Jet Propulsion Laboratory (JPL) as soon as the Deep Space Network of radio antennae received and relayed the signals from Voyager 2.

Figure 4.6 Image from Voyager 2 of Neptune, showing atmospheric streaks, the Great Dark Spot, and the Earth (for comparison). (Courtesy of NASA/JPL-Caltech.)

Nestled with my astro-buddies on the carpeted floor of a small windowless room, I witnessed the first transmitted images of Neptune *from* Neptune. As they slowly built up line by line on the television monitor, the black-and-white images showed a gray planet accented with one large dark spot and a few bright streaks (see figure 4.6). I could clearly see that Neptune resembled Jupiter with its Great Red Spot, but with less banding. Into the wee hours, my vigil continued until, six hours later, images of Neptune's largest moon, Triton, began to scroll down the TV monitor. Wow! There was nothing to prepare me for those images. Crinkled like a cantaloupe, the surface of Triton showed myriad dark streaks where icy geysers had vented. Fantastic.

Voyager 2's multi-wavelength imaging, spectroscopy, and in-situ sensing of the Neptunian system continues to provide the lion's share of what we know about this remote realm. From these data, we have learned that the planet has a significant magnetic field that, like the magnetism of Uranus, is significantly tilted and offset from the planet's center (see figure 4.5).

Neptune also resembles Uranus in size and mass, and so it probably contains a similar configuration of rock, slushy water, and liquid hydrogen within its interior, along with an atmosphere composed of gaseous hydrogen, helium and methane. Neptune appears decidedly bluer than Uranus. This could result from a higher concentration of atmospheric methane than on Uranus, or from some other unknown atmospheric effect. Neptune also radiates significantly more thermal energy than it receives from the Sun. That thermal excess helps to explain the high winds and cyclonic activity that Voyager 2 detected. Since Voyager 2's encounter with Neptune, monitoring by the Hubble Space Telescope has shown that the dark spots, bright clouds, and sundry streaks come and go. Indeed, the Great Dark Spot imaged by Voyager 2 had disappeared by the time the HST began imaging Neptune in the mid-1990s.

Currently, fourteen moons, all named after minor water deities in Greek mythology, are known to orbit Neptune. Six were discovered during the Voyager 2 encounter, while the most recent discovery was announced in 2013 – a full twenty-four years after that flyby. By far the largest of the moons is Triton. Instruments aboard Voyager 2 indicated that its surface is blanketed with various types of snow, including frozen nitrogen, water, and carbon dioxide. If the Saturnian moon Titan (which is of similar size) could be transported to the orbit of Neptune, its thick atmosphere (which is of similar composition) would likely crystallize and fall to the surface as snow. Therefore, Triton appears to be bereft of an atmosphere because it is simply too cold to sustain one.

Pluto

Until very recently, Neptune's Triton was thought to be a reasonable analogue of Pluto in terms of size, mass, and composition. However, that all changed in 2015 when the New Horizons mission flew past the icy dwarf planet, taking vivid images of its variegated surface. Pluto, which is barely bigger in projected surface area than Brazil, was revealed to sport lots of fascinating features indicating ancient and more recent cryogenic activity, completely unlike those seen on the Neptunian moon.

Cosmic debris

The Solar System plays host to a myriad of bodies that are much smaller than the eight major planets. They occupy three distinct annuli around the Sun. Most of the rocky bodies occupy the zone between the orbits of Mars and Jupiter, the so-called asteroid belt. Numbering in the hundreds of thousands, they add up to no more than 4 percent the mass of the Moon. About half of the mass in the asteroid belt is contained in the four largest objects – Ceres, Vesta, Pallas, and Hygiea. When these objects were first discovered in the early nineteenth century, they were given planetary status, thus upping the total number of planets in the Solar System. But with subsequent discoveries of yet more asteroids, prominent astronomers decided in the 1850s to demote these objects to the status of "minor planets" or simply "asteroids."

A similar downgrading of Pluto's status occurred in 2006 for exactly the same reason. Like the many asteroids that comprise the asteroid belt, Pluto and its many dwarf kinfolk beyond the orbit of Neptune occupy a shared annulus and provenance, known as the Kuiper belt. This icy deep is thought to be the original source of all comets. The main repository of comets today, however, is much farther out in the form of the Oort Cloud. We know this because many of the comets that we see passing through the inner Solar System have velocities that would place them, at their farthest extent, 5,000–50,000 AU from the Sun – well beyond the Kuiper belt.

Both asteroids and comets sometimes get perturbed into orbits that get uncomfortably close to Earth. Our planet has multiple craters from powerful impacts by these bodies. Remnants of these impactors are the iron and stony meteorites that one can find in museums of natural history.

Coda

Over the past sixty years of robotic planetary exploration, we have borne witness to incredible worlds – each with striking differences and evocative tales to tell. However, the above descriptions should be regarded as best current estimates, subject to radical updates at a moment's notice. As planetary scientist and popularizer Carl Sagan tellingly opined, "Somewhere, something incredible is waiting to be known." Indeed, to study the Solar System in the twenty-first century is to embrace these words as an operating principle.

Beyond the Solar System, a dizzying variety of stars awaits. But before we go there, we should first get ourselves better acquainted with our home star, the Sun.

5

The Sun: our star

*The sun, with all those planets revolving around it and dependent on it, can
still ripen a bunch of grapes as if it had nothing else in the universe to do.*
Galileo Galilei, *Dialogue Concerning the Two Chief World Systems*

Our star rules over the rest of the planets and shards like a god
over mortals. No wonder the Sun was the key deity in so many
pantheistic cultures of old. Today, astronomers know that it is 109
times bigger than Earth, 1.3 million times greater in volume, and
333,000 times more massive. As it shines into space, the Sun deliv-
ers to Earth's upper atmosphere 1,350 watts per square meter,
with about 2/3 of that total irradiation propagating through the
atmosphere to Earth's surface. By backtracking this radiation over
the 150 million kilometers from the Earth to the Sun, astronomers reckon that the Sun's absolute luminosity is about 4×10^{26}
watts – that's 400 trillion trillion watts! In one second, the Sun
churns out enough power to satisfy all of humankind's current
energy needs for the next 845,000 years.

Solar structure

Our best model of the Sun is as a ball of superheated gases that is
sitting upon itself. In each layer of the Sun's interior, the weight

of overlying matter is exactly countered by the outward pressures exerted by the hot gases in that layer (see figure 5.1). This perfect balancing act between inward-tending gravitation and outward-tending pressure is known as hydrostatic equilibrium. The deeper the layer, the greater the overlying weight and hence the higher the pressure that is necessary to oppose that weight. The higher pressure is provided by a combination of greater gas densities and higher temperatures. By combining the basic laws of thermodynamics with gravitational physics, astrophysicists have been able to compute the radial profiles of density and temperature that are likely to be present in the Sun (see figure 5.2).

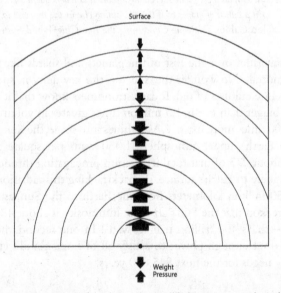

Surface

Weight
Pressure

Figure 5.1 The state of hydrostatic equilibrium ensures that the Sun neither collapses upon itself nor expands unchecked. In each layer, the weight of the overlying matter is exactly opposed by the pressure exerted by the heated gases in that layer. In the deepest layers, the gravitational and pressure forces rise to tremendous values. (Adapted from *Horizons: Exploring the Universe* by Michael Seeds, Cengage Learning [2002].)

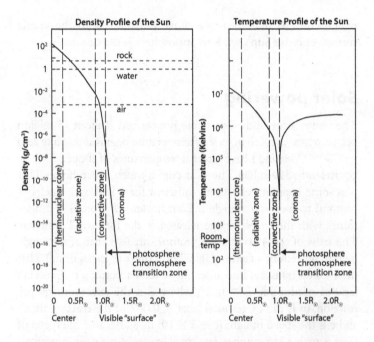

Figure 5.2 Radial profiles of the gas density and temperature inside the Sun. These profiles result from computations involving the basic laws of Newtonian gravity and classical thermodynamics. The ultimate source of power resides in the Sun's core, whose high temperature and density enable thermonuclear reactions. The energy from these reactions replenishes the energy that is radiated into space from the Sun's surface.

The density in the Sun's core exceeds that of any substance on Earth. Halfway from the core to the Sun's visible surface, the gases thin out by two orders of magnitude, achieving values close to those of water (about 1 gram per cubic centimeter). At about 90 percent of the way to the surface, the density is equivalent to that of our atmosphere at sea level. The density then takes a nosedive, ultimately plummeting by more than ten orders of magnitude (10 billion) as we move outwards from the visible surface into the Sun's tenuous atmosphere – the corona.

Solar powering

The temperature in the Sun's core is reckoned at about 15 million kelvin, where the Kelvin scale of temperature begins at absolute zero (-273 °C). You and I have internal temperatures of about 37 °C, or, equivalently, 310 kelvin. The solar core is much, much hotter. That plus some quantum effects are sufficient for 4 hydrogen nuclei (4 protons) to fuse into a single helium nucleus (2 protons + 2 neutrons), with the excess charge released in the form of 2 positrons. The mass of the 4 combining protons surpasses that of the single helium nucleus by a factor of 0.007 (just think of James Bond). That mass excess translates into an energy excess according to Einstein's famous relationship $E = m\,c^2$, where E is the excess energy per reaction, m is the excess mass (that is, 0.007 of the reacting mass), and c is the speed of light ($c = 3 \times 10^8$ m/s). Because the speed of light is such a big number, any small excess in mass can produce a huge excess in energy. As the thermonuclear reaction proceeds, the excess energy is released in the form of gamma rays and neutrinos (see figure 5.3). The Sun's thermonuclear core has enough mass to power itself in this manner for 10 billion years. Because the Solar System is no more than 5 billion years old, the Sun should be able to keep shining brightly for another 5 billion years yet.

This marvelous source of solar powering would not occur were it not for some truly weird phenomena that take place on the subatomic level. You might think that the 15-million-degree temperature in the Sun's core is pretty darn hot, but these torrid conditions are insufficient to force protons together against the mutual repulsion that their electric fields are exerting upon one another. But all is not lost, as the protons can sometimes circumvent this electric repulsion through a process called quantum

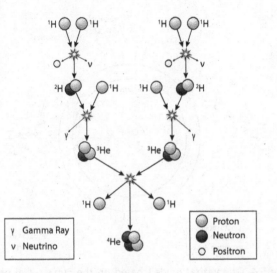

Figure 5.3 The proton–proton chain of thermonuclear reactions that is thought to power the Sun. Four protons (hydrogen nuclei) ultimately fuse into one helium nucleus containing two protons and two neutrons. The excess energy is released in the form of gamma rays and neutrinos. (Courtesy of Wikimedia Commons.)

tunneling. Simply put, if a proton bangs up against another proton enough times (say 10^{10} times), the chance of one proton getting past the electric barrier of the other proton becomes significant. The probability of quantum tunneling is just enough for thermonuclear fusion to proceed in the Sun's dense core despite its "paltry" temperature of 15 million degrees.

Energy transport

The gamma-ray photons that are released by the Sun's thermonuclear reactions ultimately heat the rest of the star and cause

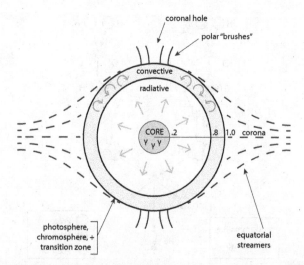

Figure 5.4 Theoretical models based on the Sun's radial profiles of internal temperature and density have revealed distinct zones, each with its own special physical properties. From the inside out, they are the thermonuclear, radiative, and convective zones. The latter's depth is constrained by helioseismic measurements, whereby the heaving Sun's surface can be sensed and analyzed in terms of favored oscillations and corresponding wavelengths. Beyond the visible surface (or photosphere), the chromosphere, transition zone, and corona demarcate solar plasmas at ever-increasing temperatures and decreasing densities.

the solar surface to radiate brightly into space. By modeling how this energy gets transported, solar physicists have come up with a set of distinct zones with unique physical properties (see figure 5.4). All of these zones consist of charged "ions" of hydrogen and helium atoms whose outer electrons have been stripped off, along with free electrons. Together, they constitute what is called a plasma. The radiative zone includes those layers which transport energy via interactions between photons of light and the atoms making up the gas. In each successive layer, increasing numbers of atoms get heated, and increasing numbers of photons are emitted by those atoms, while the energy of each re-radiated photon decreases. The result is a down-conversion of gamma-ray photons into lower-energy X-ray photons and UV photons at ever-higher layers within the Sun's interior. The process is akin to a whole lot of middlemen converting a single $100 bill into somewhat less than 10,000 pennies and pocketing their fair share. The individual interactions between photons and matter are haphazard, causing a sort of "drunkard's walk" if you were to follow any particular sequence of interactions. There is a slow outward progression, however, that follows the radial gradient of decreasing temperature in the Sun. Overall, it takes about a million years for the energy contained in the original gamma-ray photons to wend its way outward through the radiative zone.

About two-thirds of the way out from the Sun's center, the radiative zone gives way to the convective zone, where energy is transported outward by bulk boiling motions of the gaseous matter. The photosphere marks the visible surface, whence most of the photons that we can sense with our eyes originate. Just above the photosphere, the chromosphere marks the hotter, more tenuous layer whose ruby-red emission (from ionized hydrogen) is visible whenever the brilliant photosphere is eclipsed by the Moon. The diaphanous corona can be regarded as the Sun's outer atmosphere, as it extends well past Earth and even permeates the outer Solar System. The corona is more tenuous than any vacuum that can be created in the laboratory. Yet, contrary to expectations based on the classical laws of thermodynamics, it is not a low-temperature outback of the Sun. Indeed, astronomers have found the corona to be nearly as hot as the Sun's innermost core!

How the corona can be so torrid so far away from the Sun's ultimate source of power remains puzzling. Recent investigations with ground-based telescopes and spaceborne solar observatories indicate that the magnetic fields threading throughout the Sun's photosphere, chromosphere, and corona contain tremendous amounts of energy. It is this magnetic energy which astronomers believe is somehow responsible for elevating the temperature of the corona.

Solar activity

The Sun is a modestly variable star. Over the course of 11 years, the number of sunspots on the solar surface increases from a minimum wherein hardly any spots are evident to an exciting maximum 5.5 years later, then subsides back to a "quiet" minimum, thus completing a cycle that has been observed telescopically for hundreds of years. Even naked-eye observers before the time of Christ noted large spots on the Sun and their variations over time. These pre-telescopic astronomers in China, Arabia, and Europe relied upon hazy atmospheric conditions that dimmed the Sun enough for marginally "safe" viewing.

Today, we recognize that sunspots delineate relatively cool regions on the Sun that have magnetic fields thousands of times more intense than their surroundings. As the sunspots increase in number, they tie the Sun's overall magnetic field into knots of unstable magnetic energy. Eventually, the magnetic fields above the sunspots reconnect and reconfigure to lower-energy states, and in so doing, release salvos of charged particles and X-ray photons into space in the form of flares. Sometimes they loft enormous amounts of matter clear away from the Sun. These "coronal mass ejections" (CMEs) travel at speeds of 500 km/s (1.1 million miles per hour), and when directed toward Earth, blast our magnetosphere within a matter of days (see figure 5.5). The resulting auroral displays can be dazzling spectacles; the associated frying of electric power transformers, telecommunications satellites, and other electronic technologies are less appreciated consequences of these solar tirades. Pity the astronaut living on the Moon or voyaging to Mars or some relatively nearby asteroid after a CME has let go. Without Earth's magnetosphere to provide protection,

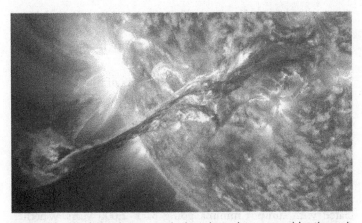

Figure 5.5 The hot solar corona is thought to be powered by the sudden reconfiguration of magnetic fields. To the left, a solar flare brightens at X-ray wavelengths, marking a particularly intense release of magnetic energy. To the right, an eruptive prominence gives rise to an enormous coronal mass ejection (CME). (Courtesy of Solar Dynamics Observatory, NASA.)

that intrepid traveler will be subject to intense radiation. This vulnerability represents one of the key challenges to long-term human exploration of the inner Solar System.

The Sun varies on other, longer timescales. Its magnetic field reverses every 11 years – returning to its original configuration every 22 years. Most likely, the 22-year magnetic field cycle and 11-year sunspot cycle are causally linked. The favored theoretical model posits that each maximum of sunspot activity scrambles the overall magnetic field to oblivion. A new global magnetic field heading in the opposite direction then arises with a subsequent decline in sunspot number. This magnetic reversal is then repeated at the next sunspot maximum. Underlying all the magnetic activity are shearing currents of conductive plasma deep within the solar interior. The relative motions between the parcels of plasma and the consequences of these motions determine the configuration of magnetic fields that emerges from the solar surface. Astrophysicists have dubbed the shearing currents the "solar dynamo," and believe that similar dynamos operate within most Sun-like stars, where convective outer layers are the norm.

On timescales of centuries, the numbers of sunspots at maximum can wax and wane. A particularly quiet period was recorded between 1645 and 1715. This so-called Maunder Minimum was associated with especially cold weather in Europe and North America. Proxies of temperature such as tree-ring thicknesses and the abundance of Carbon-14 accumulated by plants have yielded some evidence for worldwide cooling during the Maunder Minimum, and indeed during other global cooling periods associated with sunspot minima going back 2,000 years. Whether these variations in solar activity can affect terrestrial climate to such a degree remains controversial. We do know, however, that many of our technologies remain troublingly vulnerable to high solar activity. Meanwhile, the Sun's likely antics in the next decades, centuries, and millennia remain completely unknown.

6

Stars and planets beyond the Sun's domain

Blessed are the meek, for they will inherit the earth.

Matthew 5:5

Beyond the Solar System are countless other systems of stars, planets, and "debris" that migrate around the disk of our Milky Way Galaxy in orbital synchrony with the Galaxy's gravity. We have learned the most about the stellar systems that are closest to us, because we can most readily determine reliable distances to the hosting stars. As we have seen, those systems within about 100 light years of the Sun comprise what we call the Solar Neighborhood. What we have found there is a preponderance of dim, low-mass stars, along with a plethora of planets.

A cornucopia of stars

As previously shown in figure 3.9, the stars making up the innermost part of the Solar Neighborhood include the Sun (obviously), Alpha Centauri A and B, Procyon, and Sirius. Already, we can see variety in the colors of the stars that are local to us. Then there are the less recognizable stars that make up the remainder of the thirty-three stars plotted in figure 3.9. You are probably not familiar with them for a very simple reason. They are much fainter than the four listed above. Indeed, most of these stars cannot be observed without the aid of a telescope. Because they occupy the same piece of local real estate as the much brighter foursome, they must also be intrinsically less luminous. This wide variation in the luminosities and colors of stars runs rampant throughout the Solar Neighborhood. We know this because astronomers have been able to determine reliable distances, quantitative colors, and spectral types for most of these stars.

Stellar distances

As discussed in chapter 3, the best way to reckon a stellar distance is to track the star's yearly motion in the sky with respect to the background of more distant stars. This "parallactic" motion is actually the reflection of our own planet's motion around the Sun – becoming proportionately less pronounced with the greater distance of the star. By observing a star's angular parallax, and knowing the Earth's orbital "baseline" around the Sun, astronomers can triangulate the true distance to the star. For Earthbound observers, the formulation relating a star's yearly parallax (p) to its distance (d) boils down to this:

$$d \text{ (parsecs)} = 1/p \text{ (arcseconds)}$$

where p is measured in arcseconds (each of which is 1/3600 of a degree), and d is in parsecs (each of which corresponds to 3.26 light years). Indeed, the parsec unit of distance is seen to be operationally defined as a "parallax arcsecond," "parsec" being a contraction of that term. For example, Alpha Centauri A – the brightest star in the closest stellar system to the Sun – sweeps out a parallax angle of 0.747 arcseconds away from its nominal position two times each year. Its distance is therefore 1/0.747 = 1.338 parsecs, or 4.36 light years from the Sun.

To date, the distances to about 2,000 stars within 50 light years of the Sun have been accurately gauged in this way – most recently using spaceborne observatories that can more accurately measure stellar positions without confusion by atmospheric blurring.

Stellar brightnesses and luminosities

If you look up on a clear, moonless night, away from artificial lights, you can see a few thousand stars. You can also see that some stars appear much brighter than others. For example, Sirius in Canis Minor appears very bright; Tau Ceti is middling in brightness, while 61 Cygni – the first star to have its distance reliably determined – is barely visible. The Greek astronomer Hipparchus noted these variations around 135 B.C.E., and developed a system for quantifying them. According to his system, the brightest appearing stars were of "first magnitude," while the faintest stars visible to the human eye were of "sixth magnitude." Therefore, the full range of stellar brightnesses spanned magnitudes of 1–6. Sirius was then regarded as a first-magnitude star, while 61 Cygni would have weighed in with a magnitude of 5 or more.

In the nineteenth century, astronomers successfully measured the physical fluxes of light coming from the stars – where the flux denotes the received power per unit area. These fluxes, when related to the magnitude scale of Hipparchus, led astronomers to realize that the magnitude scale of brightness is logarithmic – much like the decibel scale of loudness. For every integer increase in magnitude, the stellar flux was found to decrease by a factor of 2.5. The formulation for this relation is

$$m_2 - m_1 = -2.5 \log (f_2 / f_1)$$

where m is the apparent magnitude (as observed by us), f is the measured flux in physical units of watts per square meter (W/m^2), and the minus sign accounts for the bright-to-dim nature of the magnitude scale. Using this formula, you could solve for the relative fluxes and thereby see that the difference of 5 magnitudes from the brightest- to the faintest-appearing naked-eye stars amounts to an actual flux ratio of 100. In other words, the magnitude system compresses the actual fluxes into a much smaller range of numbers. Contemporary astronomers have reworked the magnitudes accordingly, such that Sirius now has an apparent visual magnitude (m_V) of −1.5, Alpha Centauri A has a somewhat fainter magnitude of −0.1, Tau Ceti is considerably fainter at $m_V = +3.5$, and 61 Cygni strains detectability with the unaided eye at $m_V = +5.2$. According to this scale, the Sun has a whopping apparent magnitude of −26.8. Going much fainter, Barnard's Star in Ophiuchus (only six light years away) has an apparent magnitude of +9.5 and so requires a telescope to be detected.

It is one thing to quantify the flux, or apparent brightness of a star. It is quite another to know a star's intrinsic luminosity – how much energy it is radiating into space every second. To ascertain the true luminosity of a star, it is necessary to know how far away it is. That is why the Solar Neighborhood stars are so important. We know their distances from the Earth to a high degree of accuracy and so we can surmise their luminosities with confidence. The flux of light from any luminous source falls off as the square of the distance. This relation can be formulated as $f = L / (4 \pi d^2)$, where f is the measured flux, L is the intrinsic luminosity, and d is the distance. Similar inverse square laws hold for the gravitational force, electrical force, and sound intensity – as all of these "actions at a distance" involve the dilution of the pertinent effect with the increasing amount of affected area in a three-dimensional space. If you know the star's distance, you can rearrange the relation to solve for the star's luminosity such that $L = (4 \pi d^2)\, f$. Astronomers often opt to rephrase this relation in the parlance of the magnitude system. They define the absolute magnitude (M) of a star as the apparent magnitude (m) it would have if it were placed at a common distance of 10 parsecs (32.6 light years). In this way, they can rewrite the relation as:

$$m - M = 5 \log (d / 10)$$

Using this distance modulus relation and the known distances to stars, you could determine that Sirius is actually a high-luminosity star with an absolute visual magnitude (M_V) of +1.42; the Sun and Alpha Centauri A are considerably fainter at +4.83 and +4.36 magnitudes, respectively; and Barnard's Star is decidedly dim at +13.22 magnitudes.

Relating absolute magnitudes back to intrinsic luminosities involves yet another logarithmic formulation, such that

$$M_2 - M_1 = -2.5 \log (L_2 / L_1)$$

where M_2 and M_1 are the absolute magnitudes of two stars, L_2 and L_1 are their corresponding physical luminosities (in watts), and the minus sign again takes care of the "backward" nature of the magnitude scale. Solving for the luminosity ratio yields:

$$L_2 / L_1 = 10^{-0.4 (M_2 - M_1)}$$

By referring back to Sirius, you can then see that its lower absolute magnitude indicates an intrinsic luminosity that is roughly 25 times greater than that of the Sun and Alpha Centauri A. Compared to the paltry output of Barnard's Star, Sirius is blazing forth with a luminosity that is 52,000 times greater!

Stellar colors and spectral classifications

To most of us, the Sun appears yellowish. Beyond it, the brightest star in the next nearest stellar system, Alpha Centauri A, presents a similar coloration to the naked eye. By contrast, Sirius in Canis Minor appears blue-white, and Barnard's Star in Ophiuchus appears distinctly reddish when observed through a telescope. These myriad stellar colors indicate a wide range of stellar surface temperatures – with the reddest stars tracing relatively low temperatures of 2,000–3,000 kelvin, yellowish stars like the Sun tracing temperatures of 5,000–6,000 kelvin, and blue-white stars like Sirius tracing much higher temperatures of 8,000–10,000 kelvin. These interpretations of stellar colors have been confirmed by the recorded spectra of stars, whereby the light from each star is dispersed into its visible spectrum via a prism or diffraction grating (see figure 6.1).

Reddish stars like Barnard's Star have spectra rich with dips and breaks that have been correlated with the presence of particular molecules in the stellar atmospheres. The inferred existence of molecules such as titanium oxide and calcium hydroxide in the stars' atmospheres indicates surface temperatures sufficiently low to host these sorts of relatively fragile molecules – of the order of a few thousand kelvin. Yellowish stars like the Sun and Alpha Centauri A manifest no evidence for molecules in their atmospheres but show discrete dips in their spectra that denote the presence of "metallic" elements such as calcium, magnesium, and iron. The electrons in these elements are most primed to be activated by the underlying starlight when the atmospheres are at temperatures of about 4,000–6,000 kelvin. Blue-white stars like Sirius do not show molecular or "metallic" spectral features. Instead, their spectra are dominated by dips corresponding to the predicted spectrum of hydrogen. This most simple of atoms is most responsive to absorbing the underlying starlight at temperatures of about 8,000–12,000 kelvin.

By the early twentieth century, astronomers had successfully photographed the spectra of more than 200,000 stars. From visual inspection of these spectra, they devised the sequence of stellar "spectral types" that we continue to use today. From hottest to lowest stellar surface temperatures, the corresponding spectral

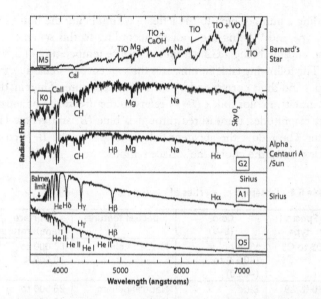

Figure 6.1 Visual spectra of nearby stars show major differences that can be attributed to the stars' differing surface temperatures. Here, the spectral characteristics of stars akin to Sirius, Alpha Centauri A, and Barnard's Star are compared. Note that "A-type" stars like Sirius are brightest at the blue (short-wavelength) end of its spectrum, while "M-type" stars like Barnard's Star are brightest at the red (long-wavelength) end – consistent with their observed colors. Moreover, the patterns of spectral absorption lines also reveal major variations that again can be ascribed to the stars' disparate surface temperatures. (Adapted from *Galaxies in the Universe* by L. S. Sparke and J. S. Gallagher, Cambridge University Press [2000], courtesy of L. S. Sparke.)

types are O, B, A, F, G, K, and M. A fun mnemonic for this coded sequence is "Oh Be A Fine Girl [or Guy], Kiss Me!" Using this codification, Sirius is an A-type star, while the Sun and Alpha Centauri A are G-type stars, and Barnard's Star is an M-type star. Further subdivision of these spectral types is accomplished by adding a number next to the letter – with 0 being the hottest of its type and 9 being the coolest. According to this scheme, the Sun ends up being a G2 star with a surface temperature of 5,800 K. The following table summarizes the sequence of stellar spectral types and their corresponding surface temperatures. Therein, the quantitative color index ($B–V$) refers to the difference of apparent magnitudes as measured through a blue (B) and yellow (V) filter. The redder the apparent color, the greater the $B–V$ color index – and the cooler the surface temperature.

Table 6.1   Observed properties of nearby stars

Spectral type	Color (B–V)	Spectral features	Surface temperature
O5 to O9	Violet-blue (−0.33) to (−0.30)	Ionized helium	60,000 to 30,000 K
B0 to B9	Blue (−0.30) to (−0.06)	Neutral helium Hydrogen	26,500 to 10,000 K
A0 to A9	Blue-white (+0.00) to (+0.19)	Hydrogen (strongest)	9,900 to 7,800 K
F0 to F9	Yellow-white (+0.31) to (0.54)	Hydrogen (weak) Ionized calcium	7,000 to 6,000 K
G0 to G9	Yellow (+0.59) to (+0.74)	Ionized calcium Ionized iron	5,900 to 5,400 K
K0 to K9	Orange (+0.82) to (+1.35)	Ionized calcium(strongest) Sodium (strongest)	5,200 to 4,000 K
M0 to M9	Red (+1.41) to (+2.00)	Titanium oxide Calcium hydroxide	3,700 to 2,700 K

(The data in this table are based on Harold L. Johnson, *Review of Astronomy and Astrophysics*, vol. 4 [1966], pp. 193–206; the online encyclopedia entry http://en.wikipedia.org/wiki/Stellar_classification; and other sources.)

It bears noting that astronomers have found other, even cooler stellar spectral types. These include the L-types, with surface temperatures of 2,500 K down to 1,300 K, and the T-types with temperatures less than 1,300 K. The L-types barely qualify as bona fide stars, where the luminous powering would come from the thermonuclear fusion of hydrogen into helium in their cores. The T-types are too cool to host hydrogen fusion and so must rely on some other source of powering – most likely the heat produced from the gravitational energy released during their formation – to explain their tepid radiation. These coolest of stellar objects are known as brown dwarfs. Because of their exceeding faintness, very little is known about them.

Putting it all together: the Hertzsprung–Russell diagram

With the distances to stars well established, astronomers can readily convert the observed stellar brightnesses (or fluxes) into intrinsic luminosities. This important stellar property, when plotted against the star's surface temperature (as determined from the star's observed color index or spectral type) yields critical clues to the types of stars that roam our Solar Neighborhood and beyond (see figure 6.2).

First devised in 1913 by Ejnar Hertzsprung of Denmark and Henry Norris Russell of the United States, the Hertzsprung–Russell (or H–R) diagram reveals distinct families among the stars. The well-populated main sequence of stars runs diagonally from the upper left to the lower right. This band hosts the vast majority of stars. Along the main sequence, the bluest and hottest stars are the most luminous, while the reddest and coolest stars are the most intrinsically dim. This can be understood as a consequence of the star's surface temperature, where the luminosity

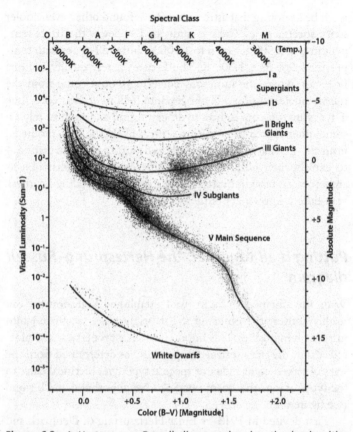

Figure 6.2 A Hertzsprung–Russell diagram showing the luminosities and surface temperatures of stars within the Solar Neighborhood and beyond. All of the stars in this diagram have distances determined from their geometric parallaxes. The diagonal main sequence is well populated, with brilliant blue stars at the upper left and dim red stars at the lower right. The giant branch has relatively fewer stars, consistent with the limited volume of space that can be fathomed with the geometric parallax technique and the relative rarity of these giant stars. The white dwarfs are sparsely populated in this diagram because of their extreme faintness, which makes them difficult to detect. (Adapted from R. Powell, *An Atlas of the Universe*.)

trends roughly as the fourth power of the temperature. The full relation can be formulated as:

$$L = \sigma \, (4 \, \pi \, R^2) \, T^4$$

which quantifies the thermal emission from the spherical surface of a hot body (idealized as a perfect radiator, or black body, where the object is in thermal and radiative equilibrium with its surroundings). The quantity $(4 \, \pi \, R^2)$ corresponds to the radiating surface area of the spherical body (in square meters), the surface temperature (T) is measured in kelvins, and σ is the Stefan–Boltzmann constant of proportionality in this formulation. When referenced to the Sun's luminosity L_{Sun}, this relation simplifies to:

$$L / L_{Sun} = (R / R_{Sun})^2 \, (T / T_{Sun})^4$$

In other words, a star's radius and surface temperature pretty much determine its luminosity. Those stars populating the main sequence vary somewhat in radius but more demonstrably in temperature. Once raised to the fourth power, the varying temperature yields tremendous variations in luminosity. Consider the main-sequence O and B stars. They are up to 100,000 times more luminous than the Sun, while the Sun is another 10,000 times more luminous than the main-sequence M stars. By contrast, the giant stars are more luminous than their main-sequence counterparts of similar temperature, because they are actually much larger. Some giants are hundreds to thousands of times larger than the Sun, for example. The same behavior holds but in the opposite way with the white dwarfs. They are much fainter than their main-sequence equivalents, because they are hundreds of times smaller. Indeed, many white dwarfs are no larger than the Earth. We will consider in chapter 12 what sorts of evolutionary processes underlie these different families of stars.

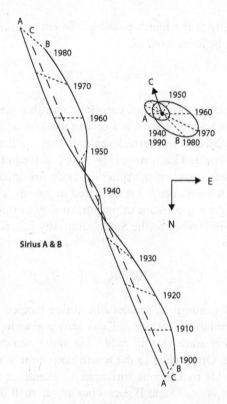

Figure 6.3 Path of the Sirius binary system in the sky. Sirius A is the bright A-type star that outshines all other stars in the naked-eye sky. Its much dimmer companion, Sirius B, is a white dwarf whose mass is roughly half that of the primary star. Each star revolves around the system's center of mass (C). (Adapted from *Introductory Astronomy & Astrophysics* by M. Zeilik and S. A. Gregory, 4th edition, Saunders College Publishing [1998].)

Stellar masses

Mass turns out to be a star's single most important physical property. It determines the star's luminosity, evolutionary path, total lifetime, and ultimate fate. Stellar masses cannot be determined by simply measuring a star's distance, temperature, and luminosity, alas. What is required is to observe the star "dancing" with another. By tracking these orbital fandangos, astronomers can invoke Newtonian gravity to derive the individual masses in the binary star system. The whole process is easiest to comprehend when both stars are visible. The Sirius A + B binary system provides an excellent example of a visual binary (see figure 6.3). Here, the bright A-type main-sequence star (Sirius A) and its much fainter white dwarf companion (Sirius B) are slowly revolving around their mutual center of mass.

Since the discovery of Sirius B in 1862, the two stars have been observed to trace out elliptical orbits that have a common period of 50.1 years. Sirius A has the smaller orbit, consistent with its greater mass. The semi-major axis of Sirius B's elliptical orbit – when referenced to Sirius A – is 7.5 arcseconds, which at a distance of 8.6 light years amounts to a mean separation of 20 AU. Using a generalized version of Kepler's Third Law, where

$$P^2 (m_A + m_B) = a^3$$

one can input the period (P) in years and semi-major axis (a) in AU, and thereby solve for the sum of the stellar masses ($m_A + m_B$) in units predicated on the Sun's mass. The resulting 3.2 solar masses can be further divvied up into individual masses according to the relative sizes of each star's orbit, such that

$$m_A / m_B = a_B / a_A$$

where the stellar masses vary inversely with the orbital radii (think of how an adult and a child are positioned on a seesaw in order to balance themselves). The size of Sirius A's orbit is about half that of Sirius B, and so its mass is twice that of the white dwarf. One ends up getting 2.15 solar masses for brilliant Sirius A and 1.05 solar masses for the exceedingly dim white dwarf Sirius B. We are now able to complete the dossier on Sirius B – a stellar object having a mass close to that of the Sun but with a luminosity that is only 2.6 percent of the Sun's output and a corresponding radius that is 119 times smaller. Clearly, Sirius B is a very different breed of star.

Compared to Sirius A + B, most binary star systems are not nearly so kind to astronomers. They are either too far away or too tightly bound to be spatially resolved. In these many instances, astronomers must glean what other information they can muster towards determining stellar masses. Those binary stars whose orbits are highly inclined to our line of sight have proved to be the most helpful. From our vantage point, the star system's total brightness will undergo periodic dips as one star eclipses the other. By monitoring these dips over time, astronomers can determine the mutual period of their orbits. Moreover, they can ascertain how closely the stellar orbits are aligned with our line of sight. If they are perfectly aligned, that means their motions towards and away from us correspond perfectly with their orbital velocities. The line-of-sight motion, in turn, can be determined from the observed Doppler shifts in wavelength of the stars' respective spectral lines. Given these special circumstances, one can determine the ratio of stellar masses as being:

$$m_2 / m_1 = v_1 / v_2 = \Delta\lambda_1 / \Delta\lambda_2$$

where the stellar mass (m) varies inversely with the star's velocity (v) and corresponding Doppler shift in wavelength ($\Delta\lambda$). Once the sizes of the orbits are obtained from the orbital period and the velocities, one can then use Kepler's Third Law to reckon the stellar masses … whew!

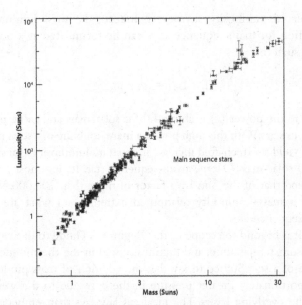

Figure 6.4 The mass–luminosity relation that has been found for main-sequence stars shows a steep dependence, with small increases in mass yielding huge increases in luminosity. (Courtesy of O. Y. Malkov, with reference to O. Y. Malkov, "Mass-Luminosity Relation of Intermediate-mass Stars," *Monthly Notices of the Royal Astronomical Society* 2007, vol. 382, pp. 1073–1086.)

Mass–luminosity–lifetime relations

Given all the demands and complexities of determining stellar masses, you might not be surprised to learn that only a couple hundred stars have been "assayed" with any decent degree of accuracy. These precious luminaries have revealed a critically important relation between a main-sequence star's luminosity and its mass (see figure 6.4). As the stellar mass increases, the observed luminosity skyrockets. Indeed, the mass–luminosity relation for main-sequence stars can be formulated as a power law, such that

$$L / L_{Sun} = (m / m_{Sun})^n$$

where the power (n) is about 4.0 for solar-mass and more ponderous stars. With this high power, a mere doubling of a star's mass will yield a sixteen-fold increase in the star's luminosity. Consider a 30 solar-mass O-type main-sequence star. Its luminosity will exceed that of the Sun by a factor of more than 100,000. Why is this mass–luminosity relation so extreme, and what are the consequences?

It is beyond the scope of this Beginner's Guide to derive the pressure, temperature, and luminosity within the thermonuclear core of a star. Suffice to say that the addition of mass produces proportionately greater pressure in the core due to the weight of the overlying layers. The ideal gas law says that temperature is directly related to pressure and so will rise with any increase in stellar mass. We have already seen that luminosity depends on the fourth power of the temperature, and so it's not hard to imagine that luminosity will depend on something close to the fourth power of the stellar mass. This feasibility argument then begs the question: "How long can a star radiate at these sorts of luminosities?"

One can approach this question by comparing the available "fuel" (which is some fraction of the star's mass [m]) with the fuel consumption rate, or "fire" (which can be identified with the star's luminosity [L]). A star's luminous lifetime can then be estimated by dividing the available "fuel" by the "fire." When referenced to the Sun, the star's approximate lifetime (τ) becomes:

$$\tau / \tau_{Sun} = (m / m_{Sun}) / (L / L_{Sun})$$

If the luminosity (L) is replaced by its mass equivalent, according to the mass–luminosity relation, this relation for stellar lifetime reduces to:

$$\tau / \tau_{Sun} = (m / m_{Sun})^{-3.0}$$

Here, we have yet another extreme power law, this time with increased mass producing a major *decrease* in the star's lifetime. The Sun's total thermonuclear lifetime can be calculated from the mass of the Sun ($m_{Sun} = 2 \times 10^{30}$ kilograms), the mass of the thermonuclear core (about 10 percent of the total mass), the fraction of that mass that undergoes thermonuclear fusion (about 0.007 the core's mass), the corresponding energy that is released over the Sun's lifetime according to Albert Einstein's famous equation $E = m\ c^2$, and the Sun's measured luminosity ($L_{Sun} = 4 \times 10^{26}$ watts). All these considerations boil down to a total thermonuclear lifetime of:

$$\tau_{Sun} = E_{Sun} / L_{Sun} = (0.007 \times 0.1 \times m_{Sun})\ c^2 / L_{Sun}$$

Plugging in the values for m_{Sun} and L_{Sun} yields an estimated thermonuclear lifetime for the Sun of roughly 10 billion years (10^{10} years), so we're currently near the halfway mark. If this ballpark figure is plugged into the prior relation for relative lifetimes, the estimated lifetime of a main-sequence star becomes:

$$\tau = 10^{10}\ (m / m_{Sun})^{-3.0}\ \text{years}$$

which is a reasonable approximation to what the more sophisticated stellar models indicate (see table 6.2). For example, a 10 solar-mass B3 main-sequence star has a predicted lifetime of only 10 million years, while a 0.7 solar-mass K5 star should last about 30 billion years – much longer than the present 13.8-billion-year age of the Universe itself.

Table 6.2 Physical properties of nearby main-sequence stars

Spectral type	Mass	Luminosity	Lifetime	Relative abundance
O	>16 m_{Sun}	>30,000 L_{Sun}	<5.0 Myr	0.00003%
B	16 to 2.1 m_{Sun}	30,000 to 25 L_{Sun}	5.0 to 840 Myr	0.13%
A	2.1 to 1.4 m_{Sun}	25 to 5.0 L_{Sun}	0.84 to 2.8 Gyr	0.6%
F	1.4 to 1.0 m_{Sun}	5.0 to 1.5 L_{Sun}	2.8 to 6.9 Gyr	3.0%
G	1.0 to 0.8 m_{Sun}	1.5 to 0.6 L_{Sun}	6.9 to 13 Gyr	7.6%
K	0.8 to 0.5 m_{Sun}	0.6 to 0.1 L_{Sun}	13 to 56 Gyr	12.1%
M	0.5 to 0.1 m_{Sun}	< 0.1 L_{Sun}	>56 Gyr	76.5%

(Masses, luminosities, and relative numbers refer to values in en.wikipedia.org/wiki/Stellar_classification. Lifetimes in millions of years [Myr] and billions of years [Gyr] are calculated from $\tau = 10^{10}$ (m / m_{Sun}) / (L / L_{Sun}) years. There are four B-type stars within 100 light years of the Sun. The nearest O-type star, Zeta Ophiuchi, is about 400 light years away – well beyond the limits of the Solar Neighborhood considered here.)

What, then, will become of the Solar Neighborhood? In another billion years or so, Sirius will be gone. After 5 billion years, the Sun will call it quits. Meanwhile, every red, dim, M-type star that was ever made – going back to the birth of the Milky Way Galaxy itself, some 12 billion years ago – will continue to abide and so characterize our little patch of galactic real estate. Star-forming clouds of gas and dust may come and go, but it will be the puny M dwarfs that prevail.

Exoplanets galore!

The news on planets beyond the solar domain keeps getting ever more amazing. Recently, the *Extrasolar Planets Encyclopaedia* listed more than 4,500 confirmed exoplanets in more than 3,200 systems. By the time you read this section, there could be lots more planets added to the list. All of these scientific discoveries have occurred since 1992, when the first planet beyond the Sun was found orbiting a neutron star, of all things. Before then, there were decades of false alarms and dashed hopes. I distinctly remember teaching astronomy classes in the 1980s and enthusiastically announcing the first exoplanetary discovery, only to rescind the news in subsequent classes. Consider the history of exoplanetary discoveries, and you can begin to appreciate the excitement associated with this rapidly expanding field of astronomy.

Before 2010, the vast majority of discoveries were made by closely monitoring the spectra of the host stars. By obtaining high-resolution spectra and measuring the minute Doppler shifts in wavelength of the stars' spectral lines, astronomers began to find instances of periodic wobbling of the stars. The Doppler shifts amounted to less than 1/10 million of the nominal wavelengths, but that was enough to infer stellar wobbling of a few meters per second (walking speed). These sorts of stellar gyrations indicated the gravitating presence of one or more nearby planets. The advantage of the Doppler-shift method for finding exoplanets is that the astronomers can determine both the orbital period and the orbital velocity of the planet (accounting for the orbital inclination), which then can yield the gravitating mass of any planet, along with the distance of the planet from its host star.

Since 2010, most of the planetary discoveries have come from the Kepler Space Telescope. Launched in March 2009, the solar orbiter has yielded a bonanza of planets – including some no bigger than Earth. The Kepler telescope was purposely tasked to continuously image one patch of sky that straddles the constellations of Cygnus the Swan and Lyra the Harp. Within this patch are approximately 150,000 stars whose light Kepler could monitor with uncanny precision. Besides tracking a star's natural variability in brightness, Kepler was sensitive to any temporal dips in light that might be produced by a planet transiting in front of the star. Using these light profiles, astronomers have been able to determine both the orbital periods and the sizes of the transiting exoplanets.

In order to obtain the planetary masses, however, it is necessary to follow up with spectroscopic observations that reveal how fast the planets are orbiting their host stars. The findings from Kepler along with a suite of dedicated ground-based telescopes have together revolutionized what we know about extrasolar planets.

If we restrict ourselves to those within a hundred light years of us, we are privy to about 650 known planets. These include planets around our next-door neighbors, Proxima Centauri and Alpha Centauri B, as well as other stars visible to the naked eye. For some systems, we've even managed to obtain images of planets orbiting their host stars.

The actual numbers of planets by type remain uncertain. That is because the most productive methods for finding planets – the "wobble" and "transit" techniques – are biased toward detecting large planets close to their host stars. The wobble technique depends on the planet gravitationally perturbing the hosting star. More massive planets and/or planets closer to their star will produce the most detectable wobbles. The transit technique depends

PROXIMA CENTAURI

On August 24, 2016, European astronomers reported the discovery of a planet in orbit around the nearest star to our Solar System, Proxima Centauri. This dim red dwarf star is an outlying member of the Alpha Centauri triple star system. Evidence for the planet was obtained using the "stellar wobbling" method, whereby the central star's spectrum was observed to shift in wavelength ever so slightly with a period of 11.2 days. The planet, Proxima Centauri b, is estimated to have a mass of at least 1.27 Earths and an orbit with a semi-major axis of only 0.05 AU. Despite its startlingly close proximity to its host star, the planet is thought to have a surface temperature similar to that of Mars (–39 °C), due to the star's very low temperature and luminosity. Perhaps the most important part of this is that the Proxima Centauri and Alpha Centauri B planetary systems, being the closest to our Solar System, represent the best exoplanet prospects for us to someday explore with robotic spacecraft.

on the planet significantly obscuring the light from the background star. The bigger the planet, the deeper will be the resulting dip in total brightness during transit. And the closer the planet is to its star, the greater the probability that it will transit the star as seen by us. For these reasons, the current harvest of exoplanets shows an excess of Jupiter-size planets well within the equivalent of Earth's orbit.

Despite these observational biases, astronomers have found a significant number of less massive planets that are far enough away from their stars to conceivably host surface water in liquid form. Moreover, they have been able to surmise from the planets' derived masses and sizes the overall constitution of these planets – whether they are primarily gaseous, liquid/ice, or rocky. Already, we have poster children for "super-Earths" containing profound oceans that are situated within the so-called habitable zone, along with a handful of Earthlike rocky planets. Meanwhile, the numbers of stars with detected planets seem to indicate that almost all stars should be hosting at least one planet. Given the number of stars that may be roaming the disk of our Galaxy, that estimate would imply between 100 and 400 billion planets out there.

The Kepler mission scrutinized fairly distant stars within a small patch of sky towards the constellations of Cygnus and Lyra, as described above. Its follow-on mission, TESS (Transiting Exoplanet Survey Satellite), which launched in April 2018, is surveying relatively nearby stars but across the entire sky. This mission has already begun to harvest several new exoplanetary systems which, being closer, can be more effectively diagnosed through spectroscopic observations with ground-based telescopes and, now, the James Webb Space Telescope.

To learn about stars and their planets in a more comprehensive way, we will need to spread our horizons well beyond what we have surveyed in this chapter and so consider the full sweep of the Milky Way Galaxy. The following chapter will take on that challenge – introducing our Galaxy's sundry stellar, nebular, and dark contents along with its exquisite architecture and dynamics.

7

The Milky Way Galaxy

I have observed the nature and the material of the Milky Way. With the aid of the telescope this has been scrutinized so directly and with such ocular certainty that all the disputes which have vexed philosophers through so many ages have been resolved, and we are at last freed from wordy debates about it. The galaxy is, in fact, nothing but a congeries of innumerable stars grouped together in clusters. Upon whatever part of it the telescope is directed, a vast crowd of stars is immediately presented to view. Many of them are rather large and quite bright, while the number of smaller ones is quite beyond calculation.

Galileo Galilei, *The Starry Messenger*

The Milky Way has transfixed and entranced humans since deep prehistoric times. On clear, moonless nights, our forebears bore rapt witness to this ghostly band of ethereal light. Indeed, the stunning sight demanded their attention. Extending from horizon to horizon, its patchy raiment gave structure to our ancestors' myriad myths of origin. These included the Milky Way as a pathway for immortals to travel, as a river that divided two star-struck lovers, as a parade of dark creatures amid a luminous background, and as the milk spilt by a nursing goddess.

Nowadays, it is difficult to recreate the visceral sensations that our distant kin must have experienced upon viewing the Milky Way. Artificial lighting has all but blotted out the more sublime apparitions of the night sky. Even from my seaside home in Rockport, Massachusetts, where ocean surrounds me on three sides, I must contend with light pollution from the nearest cities to the south.

Delineating the Galaxy

Beginning with Galileo Galilei's "congeries of stars," we have since learned that the Milky Way represents our "insider's" view of a flattened stellar system. The basic layout of the Milky Way Galaxy – and our particular vantage point within it – is shown in figure 7.1. From the inside out, our home Galaxy has a bulge–disk–halo architecture. The Sun and Solar System reside in the disk component, about two-thirds of the way out from the central bulge. The flattened disk contains almost all of the stars that are visible in the night sky, plus a whole lot more that is invisible to us.

The obscuring effects of interstellar dust in the disk prevent us from seeing more than a few thousand light years in toward the Galactic center – another 27,000 light years farther away. This dust consists of microscopic grains of icy silicates and graphites that co-dwell with gobs of hydrogen and helium gas in the form of giant, dark, star-forming clouds. Attenuating the light beyond them, these clouds are responsible for the Milky Way's patchy appearance. They and the newborn stars that they spawn are likely arranged along spiral arms – though astronomers remain uncertain as to the actual disposition of the spiral arm pattern. Perambulating around the Galactic center on orbits lasting hundreds of millions of years, these ponderous clouds represent our Galaxy's future.

The bulge component is noticeable to the unaided eye, as it extends above and below the murky disk in the directions of Sagittarius, Ophiuchus, and Scorpius. With the assistance of binoculars, you can see "star clouds" that actually delineate windows in the dusty haze through which the stars of the bulge are visible.

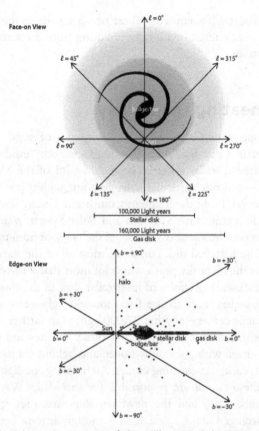

Figure 7.1   Schematic cartoons of the Milky Way Galaxy as seen from well beyond its extremities. These views are based on a combination of observations and subsequent explanatory models. *Top:* Face-on perspective of the Milky Way, featuring the central bulge/bar of stars, surrounding stellar and gaseous disk, and spiral arms in the disk that trace recent star-forming activity. The arrangement of Sun-based Galactic longitudes (l) is also presented. *Bottom:* Edge-on perspective of the Milky Way, featuring the central bulge/bar, thin stellar and gaseous disk, and halo that contains globular star clusters (shown) and ponderous amounts of dark matter (not shown). The arrangement of Sun-based Galactic latitudes (b) is also presented. (From *The Milky Way* by W. H. Waller.)

Figure 7.2   The Milky Way Galaxy, as viewed from far away, would likely resemble the barred-spiral galaxy Messier 109, shown here. (Courtesy of S. Swanson and A. Block, National Optical Astronomy Observatories, Associated Universities for Research in Astronomy, National Science Foundation.)

Like the bulges of other galaxies, our bulge contains mostly yellow, orange, and red stars. This limited assortment of hues contrasts with the full range of stellar colors that are visible in the disk. It appears that the bulge contains mostly old stars of low surface temperature, while the disk continues to host young, hot, blue stars along with a rich legacy of older, ruddier stars.

The shape of the bulge remains controversial. At least some part of it appears decidedly oblong, in the form of a bar that is inclined by about 45 degrees to our particular line of sight. Other galaxies sporting thin disks and spiral arms contain these sorts of central bars (see chapter 8). If viewed from afar, our Galaxy would be classified as a barred-spiral galaxy. The galaxy Messier 109, located 46 million light years away in the constellation of Ursa Major, provides a close analogy to what our Galaxy might look like.

The halo component would be mostly invisible were it not for the globular star clusters that swarm within it. These tight groupings of thousands to millions of stars go back to the earliest times – when the Milky Way Galaxy was still forming from the chaos of colliding and merging matter that followed the hot big bang. Only long-lived low-mass stars populate the globular clusters today. The more massive short-lived stars expired long ago, leaving behind relics of what the clusters must have looked like. Even so, they look fabulous – like bejeweled brooches on a black velvet background. A decent amateur telescope will give you impressive views of these stellar baubles.

The halo and bulge collectively comprise the Galaxy's spheroidal component. Both venues contain stars that are old and lacking in heavy elements (though the bulge also contains metal-rich stars of varying ages). Excessively fast motions of the stars and gas in the disk component have led astronomers to surmise that there is something else populating the spheroidal component. Overwhelming in its mass, this ineffable something is responsible for gravitationally restraining the disk's zippy stars and gas, and so keeping the Galaxy from dissipating into nothingness. Astronomers have yet to get a glimpse of the gravitating material and so they have dubbed it "dark matter." Searches for the sundry exotic particles that have been proposed to comprise the dark matter continue today. So far, no convincing candidates have been found (see chapter 13). Despite this lack of success, astronomers continue to regard galaxies as consisting mostly of dark-matter halos along with a smattering of ordinary matter within their swarming bulges and swirling disks that light up the dark and – in the case of the Milky Way – provide for our inheritance.

Galactic constituents

Much of what we know about the Milky Way's disk, bulge, and halo components has come from the painstaking task of determining accurate distances. In chapters 3 and 6, the method of geometric parallax was cited as the primary technique for determining stellar distances throughout the Solar Neighborhood. As we saw, this method depends on Earth's orbit around the Sun as the baseline for triangulating distances to nearby stars. To effectively explore beyond the Solar Neighborhood, however, astronomers have had to enlist other, more far-reaching methods. On the scale of the Milky Way Galaxy, star clusters have proved to provide the best milestones.

Star clusters

Some star clusters can be viewed with the naked eye. These include the Pleiades cluster (the "Seven Sisters") located in Taurus the Bull (see figure 7.3), the Hyades cluster which makes up the snout of Taurus, and the double cluster of h and chi Persei in Perseus the Hero. If you tour the Milky Way with binoculars, you will find many other smudgy objects that upon closer inspection with a telescope can be resolved into clusters of stars. The adjoining constellations of Gemini the Twins and Auriga the Charioteer, for example, have many star clusters that are readily observable through binoculars and small telescopes. Like the other 110 fuzzy objects that were first catalogued in 1784 by Charles Messier, these star clusters have Messier designations – the most prominent in Gemini and Auriga being M35, M36, M37, and M38.

The key benefit of using star clusters to gauge distances is the fact that each of the stars in a cluster is situated at pretty much the same distance from us. That means the variations in brightness from star to star can be regarded as actual differences in their respective luminosities. Through careful measurement of the light

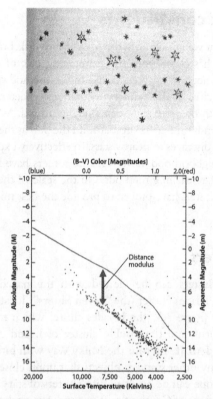

Figure 7.3 *Top:* Galileo Galilei's mapping of stars making up the Pleiades star cluster (M45). *Bottom:* A modern comparison of the stellar colors (*B–V*) and apparent magnitudes (*m*) in M45 with respect to the calibrated main sequence of colors and absolute magnitudes (*M*), as represented by the sinuous line. The difference between apparent and absolute magnitudes (*m – M*) is known as the distance modulus, from which the distance to the cluster can be calculated. Here, the distance modulus is 5.6 magnitudes, which yields a distance of 132 parsecs (475 light years). (*Top: Siderius Nuncius* [Starry Messenger] by G. Galilei, Venice, Italy, 1610; courtesy of History of Science Collections, University of Oklahoma Libraries. *Bottom:* Adapted from interactive cluster-fitting program of K. Lee, University of Nebraska, Lincoln, available at http://astro.unl.edu/naap/distance/animations/clusterFittingExplorer.html)

from each star in a cluster, astronomers have been able to construct color–magnitude diagrams (CMDs) which reveal the cluster's main sequence (see figure 7.3). Upon comparison with the main sequence from a fully calibrated set of nearby stars, they can then determine the cluster's distance modulus ($m - M$) and corresponding distance (see caption to figure 7.3). This important method of determining stellar distances is known as main-sequence fitting.

Beginning with nearby clusters such as the Pleiades and Hyades, astronomers have been able to employ similar tactics to fathom distances to star clusters throughout a significant fraction of the disk and into the halo as well. These distances have provided the crucial "skeleton" onto which astronomers have elaborated models that flesh out the overall structure of the Milky Way (as depicted in figure 7.1).

The star clusters are fascinating in and of themselves, as some sport a full range of blue, yellow, and red stellar colors, while others contain only yellow and red stars. Through careful analyses of the clusters' color–magnitude diagrams, astronomers have been able to show that the differing stellar populations can be diagnosed in terms of their corresponding ages (see figure 7.4). The key to these sorts of diagnoses is to find the color and luminosity where the cluster's main sequence terminates. This so-called main-sequence turnoff corresponds to the stars which are just now ending their normal lives and becoming red giants. A cluster with a main-sequence turnoff that includes hot blue stars is relatively young, as the massive blue stars have yet to evolve away from the main sequence and ultimately expire. A cluster with a turnoff populated by yellow or red stars, however, has already lost its hotter, bluer, brighter, and more massive stars. Using a cluster's observed main-sequence turnoff and the stellar color–luminosity–lifetime relations discussed in chapter 6, astronomers can read off the main-sequence lifetimes of the stars making this transition and thereby reckon the age of the cluster as a whole. In so doing, they have discovered that most star clusters in the disk are no more than a few billion years old.

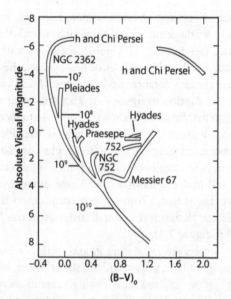

Figure 7.4 Composite color–magnitude diagram of prominent star clusters in the Galactic disk. The double cluster of h and chi Persei has a fully populated main sequence of stars, including hot, short-lived B-type stars, thus indicating a young age of about 10^7 years. By contrast, the Pleiades cluster's main sequence no longer includes such hot, high-luminosity stars. Their absence betrays a correspondingly older age of 10^8 years. The Hyades cluster has an even more truncated main sequence, consistent with an age of 10^9 years. (Adapted from *The Milky Way* by B. J. Bok and P. F. Bok, 5th edition, Harvard University Press [1981], with reference to H. L. Johnson and A. R. Sandage, *Astrophysical Journal*, vol. 121 [1955], pp. 616–627.)

Conditions in the disk appear to be hostile to the integrity of star clusters. Dissipation of their natal clouds, along with close approaches by other star clusters and gas clouds, are likely responsible for having gravitationally perturbed and ultimately dispersed most of these stellar collectives. Many astronomers regard the Sun and Solar System as having been born as part of a substantial cluster. Over the 4.6 billion years since its formation, however, this cluster has come apart. We now appear to be one of many orphaned solar systems that roam the disk – living out our existence as a stray among other strays.

Circumstances in the halo are far more amenable to the sustenance of the globular star clusters that reside there. By swarming in orbits that mostly stay away from the congested disk, globular clusters in the halo can abide much longer without serious provocation. Using similar dating techniques, astronomers have found that globular clusters have decidedly ancient ages of 11–13 billion years. These cluster ages provide a lower limit for the age of the Milky Way itself and for the entire galaxian Universe.

Super stars

As discussed in chapter 6, our Solar Neighborhood is dominated by dim red stars that can only be spied with a telescope. Conversely, our naked-eye views of the night sky are dominated by super stars – luminaries so brilliant that they can be viewed from great distances. Consider the constellation of Orion the Hunter. The prominent stars in this distinctive constellation are all extraordinarily distant and powerful. The brightest of these, Rigel (Beta Orionis A), is a blue supergiant star an estimated 900 light years away. It is 79 times larger than the Sun and radiates 120,000 times more brilliantly. Then there is Betelgeuse (Alpha Orionis), a red supergiant reckoned to be 520 light years distant. Though considerably cooler than Rigel, it is so huge that its surface puts out the same equivalent 120,000 Suns worth of luminosity at longer wavelengths. Direct imaging of this super star has yielded a size that spans a thousand Suns. Indeed, if it were to replace the Sun at the center of the Solar System, Betelgeuse would engulf all planets out to Jupiter. You can gain an appreciation for the range of sizes among the stars by examining figure 7.5.

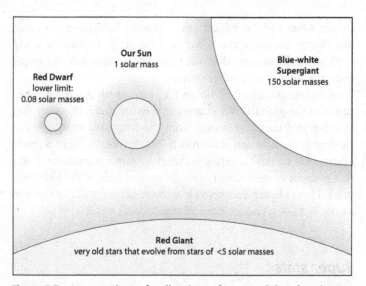

Figure 7.5 A comparison of stellar sizes – from a red dwarf to the Sun and much larger giant stars. (Courtesy of the European Space Agency, Hubble Space Telescope.)

Unlike the situation with star clusters, the distances, sizes, and luminosities of these individual super stars are far less certain. Through detailed analyses of the stellar spectra, astronomers have found tell-tale signs relating to the degree of stellar bloating. For example, a relatively compact main-sequence star has a dense radiating atmosphere whose pressures produce higher-velocity swarming of the atoms and corresponding Doppler broadening of the spectral absorption lines. By contrast, a red supergiant like Betelgeuse possesses a more tenuous atmosphere of low pressure which yields significantly narrower spectral absorption lines than its main-sequence equivalent. Using these sorts of spectral characteristics, astronomers have devised luminosity classes associated with the main-sequence, giant, and supergiant stars. These go from luminosity class I for a supergiant like Betelgeuse to luminosity class V for a main-sequence star like the Sun. Armed with these classifications, and fortified by additional information from the observed properties of equivalent-type stars in clusters, astronomers have been able to estimate the radiating sizes, luminosities, and distances of individual giant and supergiant stars with uncertainties of about 25 percent (see figure 7.6).

Considerable improvements upon this admittedly shaky situation are progressing with the Gaia astrometry mission. Launched in December 2013, this European spacecraft aims to measure the geometric parallaxes and so determine the distances of 20 million stars to better than 1 percent accuracy. It will also reckon to better than 10 percent accuracy the distances of another 200 million stars – as far away as the Galactic center. When these distances are combined with the velocity measurements that Gaia will make, astronomers will finally be able to map out the three-dimensional stellar architecture of the Milky Way – and determine how all these stars are moving about the Galaxy. As if that isn't enough, Gaia is also expected to find thousands of new exoplanetary systems.

Variable stars

In chapter 5, we saw that our Sun is variable in its luminous output – and that this variability can have potentially serious consequences here on Earth. The Sun's average variation in luminosity is about 1 part in 1,000 (0.1 percent), with the ultraviolet portion of the Sun's spectrum being about fifteen times more variable (at 1.5 percent). Now consider the bona-fide pulsating variable stars that roam the Milky Way. First noted in the 1600s, these so-called intrinsic variable stars come in all colors, sizes, and luminosities (see figure 7.7). (Eclipsing binary stars also show periodic changes in brightness, as one passes in front of its partner. These systems

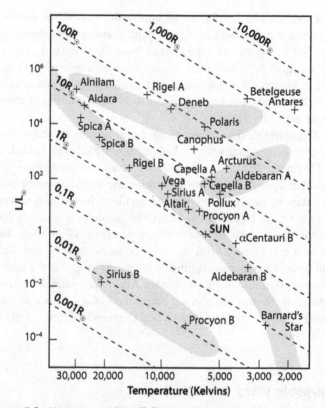

Figure 7.6 Hertzsprung–Russell diagram showing the full range of stellar luminosities and temperatures (compare with Figure 6.4). Many of the brightest naked-eye stars are evident as well as some of the most intrinsically faint stars. Lines of constant radius show that stars along the main sequence vary somewhat in size, while the respective sizes of giant and white dwarf stars are decidedly disparate. (Adapted from *Horizons: Exploring the Universe* by Michael Seeds, Wadsworth/Cengage Learning, Inc. [2002].)

are essential to determining stellar masses [see chapter 6], but will not be considered here.)

The levels of variability in the intrinsic variable stars can range from a few tens or hundreds of percent in the RR Lyrae variables,

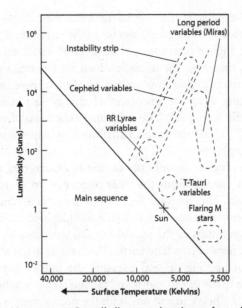

Figure 7.7 Hertzsprung–Russell diagram showing a few of the many differing types of intrinsic variable stars. The instability strip contains stars that pulsate at resonant frequencies in response to internal provocations. (Adapted from *Discovering the Essential Universe* by N. F. Comins, 3rd Edition, W. H. Freeman [2006].)

to several million percent in the Mira variables. Moreover, the periods of variation depend on the stellar types. The RR Lyrae stars, for example, are A-type giants that pulsate with periods of several hours to a few days. The Cepheid variables are F- to K-type supergiants that periodically vary over days to months, respectively. The Mira variables – also known as long period variables (LPVs) – are a type of red giant that is approaching the end of its life. These unstable stars take months to years to slowly but drastically change their luminous output.

The menagerie of intrinsic variable stars is truly staggering in its diversity. At the red, low-luminosity end, there are the flaring M-type stars. The outbursts from these low-mass stars are associated with magnetic storms in the stars' atmospheres. At somewhat higher luminosities, the T-Tauri variables are new-born stars that have yet to settle down on the main sequence. Their relatively higher luminosities are due to these infant stars still radiating away the gravitational energy that was released during their formation. The wild aperiodic variability charac-teristic of T Tauri stars is again attributed to magnetic tirades in their atmospheres.

At even higher luminosities on the H–R diagram, the insta-bility strip extends from the main sequence up to the super-giant branch. This diagonal locus of stellar surface temperatures and luminosities hosts the RR Lyrae and Cepheid variables. The internal conditions in the stars along this strip lead to a type of dynamical instability, where particular layers undergo alternating episodes of bottling up and releasing the radiant energy from below. The result of this radiative "valving" is a star whose outer layers literally heave in and out, with the surface temperature and luminosity periodically increasing and decreasing in lockstep.

Commonly found in globular clusters and other parts of the Galaxy's older spheroidal component, the RR Lyrae variables pul-sate about a well-established luminosity of about 50 Suns (M_V = 0.6 mag). This characteristic average luminosity provides what is known as a standard candle, by which astronomers can gauge the distances of globular clusters and other Galactic venues that contain these stars. For example, observations of the many RR Lyrae stars resident in the Galactic bulge have shown that the bulge – and the Galactic center that it engulfs – is around 27,000 light years away.

Two kinds of Cepheids have been discovered along the insta-bility strip. The "classic," or Type I Cepheids, dwell in the disk, while the lower-luminosity Type II Cepheids typically reside in much older globular clusters belonging to the halo. A criti-cally important relation has been found for the supergiant Type I Cepheids that has enabled astronomers to determine distances to stars well beyond the Milky Way. By observing Cepheid variables in the Small Magellanic Cloud (one of the Milky Way's closest companion galaxies), astronomers in the early twentieth century discovered a tight correlation, such that the higher-luminosity Cepheids hosted longer-period oscillations. This period–luminosity relation, once calibrated by nearby Cepheids in the disk of our Galaxy, has since allowed us to determine distances to any galaxy whose Cepheids can be resolved by our telescopes.

By monitoring a distant Cepheid's light output over a few days to several weeks, an astronomer can establish the star's period of variability. The period can then be translated into an intrinsic luminosity for the star, which in turn can be compared with the star's apparent brightness for the purpose of calculating how far away it is. First discovered by Henrietta Leavitt in 1912, and used by Harlow Shapley in the 1920s to establish the spatial distribution of globular clusters in the Galaxy, the Cepheids' period–luminosity relation has since been employed in the fathoming of distances to galaxies tens of millions of light years away.

At highest luminosities, the luminous blue variable stars delineate the upper limit of stellar integrity. The flux of photons from the surfaces of these newborn screamers is so great as to completely destabilize the outer layers and blow them away in the form of titanic winds. Eta Carinae, located 7,500 light years away in the southern constellation of Carina the Keel, is currently undergoing these sorts of dramatic eruptions. Its most powerful recorded outburst occurred in 1843, when it was the second-brightest star in the sky (after Sirius), despite its great distance. Since then, Eta Carinae has produced a small bipolar nebula of outrushing gases in the form of an exquisite hourglass (see en.wikipedia.org/wiki/Eta_Carinae). When it will erupt again – or explode entirely – is anyone's guess.

Gaseous nebulae

Cruising around the Galactic disk are about 6,000 humongous clouds of molecular gas and dust. Each cloud extends for tens to hundreds of light years and contains upwards of a million Suns' worth of nebular matter – all at temperatures of just a few degrees above absolute zero (−273 °C). These cold, dark clouds consist of about 73 percent molecular hydrogen, 25 percent atomic helium, and a smattering of other molecules such as carbon monoxide and formaldehyde, along with a haze of microscopic dust grains. You can see some of these dusty clouds as "dark nebulae" in silhouette against the glow of the stellar Milky Way. The Great Rift that cleaves the constellations of Aquila and Cygnus, the Pipe nebula in Ophiuchus, and the Coalsack in Crux are notable examples of relatively nearby molecular clouds obscuring the light from more distant stars. The Quechua Indians of Andean Peru interpreted these myriad dark features in the luminous Milky Way as a fox, llama, partridge, and various mythic creatures – thus reversing the more typical way of looking at the Milky Way as a glowing construct on a black background.

Nowadays, astronomers study the dark nebulae by observing the light that their molecules radiate. Given the cryogenic temperatures, most of the light is emitted in the low-energy microwave part of the electromagnetic spectrum. Molecular hydrogen, though the most abundant molecule by far, is a terribly weak radiator, except when excited by strong ultraviolet irradiation or interstellar shock waves. Carbon monoxide, however, readily radiates whatever energy it has acquired from stars, cosmic rays, or even the cosmic microwave background. Astronomers have exploited carbon monoxide's strong emissivity to map the spatial distribution of the molecular clouds throughout the disk of the Galaxy. They have found a strong preference for the clouds to orbit within an annulus that extends from 11,000 to 23,000 light years from the Galactic center. They have also found some indications that the clouds are arrayed along spiral arms, though the exact number and shape of the arms remain controversial.

Detailed observations of the largest molecular clouds in the light of carbon monoxide and other emitting molecules have revealed them to be enormous and complex in structure – each one spanning hundreds of light years and containing up to a million solar masses' worth of molecular hydrogen. Indeed, the few thousand giant molecular clouds known to be lumbering about the Galactic center comprise the largest objects of any kind that can be found in the Milky Way Galaxy. Inside these behemoths, wonderful transformations are taking place, whereby further condensations of the molecular gas ultimately lead to the birth of stars in clusters. These clustered stars exert energetic feedback upon their natal clouds, thus giving rise to all sorts of architectural changes and emissive phenomena.

If you go online and search under "Milky Way," you will no doubt find a wonderful assortment of images depicting our Galaxy and the many evocatively dark nebulae that ripple through the stellar tableaus. You may also see some smaller, brighter gaseous regions glowing with roseate hues. You can find stunning close-ups of these so-called emission nebulae by searching under "Nebulae."

Emission nebulae come in three basic types – HII regions, planetary nebulae, and supernova remnants. HII regions are parts of molecular clouds that have recently formed hundreds to thousands of stars, organized in clusters. The most massive of these stars are incredibly hot and powerful. Their ultraviolet light, in particular, is responsible for breaking apart the gaseous molecules into atoms and then stripping the atoms of their outermost electrons. This process of photo-ionization produces a plasma of positively charged ions and negatively charged electrons which together have a characteristic temperature of several thousand degrees. The sobriquet of "HII" refers to the singly ionized state of hydrogen, whereas "HI" denotes neutral atomic hydrogen. Having only one electron, hydrogen can exist only in the forms of HI and – once ionized – HII.

Photo-ionized hydrogen radiates at specific wavelengths whenever an electron is recaptured by a hydrogen ion and then jumps down a cascade of energy levels. The electron's jump from the third to the second quantized energy level produces the ruby-red glow that characterizes many deep images of emission nebulae. The visible part of the hydrogen spectrum also sports fainter emission lines of teal, blue, and violet hues. Other atoms have more electrons available to be ionized and so can have a greater variety of ionization states and corresponding electronic transitions. In HII regions, the ions of oxygen (OII and OIII), nitrogen (NII), and sulfur (SII and SIII) glow with an assortment of red, green, blue, and violet colors that are associated with specific jumps in the quantized energy states of their remaining outermost electrons. Long-exposure images of these HII regions often reveal multi-hued cavities and bubbles of irradiated and photo-evaporating gas. Some of these "working surfaces" sport protuberances directed inward towards the energizing stars; these sculptures of gas and dust are akin to the rain-eroded rocky "hoodoos" that jut above their surroundings in Bryce Canyon and other weathered lands. Observations at infrared wavelengths have shown these complex regions to be hosting lots of complex organic molecules. Indeed, we are witness to the beginnings of pre-biotic chemistry in these energized ecosystems. The most famous of the HII regions is the Orion Nebula (M42), located in the sword that dangles (rather dangerously) below Orion's belt. Any decent amateur telescope will give you good views of the nebula and its exciting stars. Other well-loved HII regions include the Eagle nebula (M16) in Sagittarius (home to the iconic "pillars of creation") and the Carina nebula in its southern namesake constellation (the brightest emission nebula in the sky and home to the eruptive star Eta Carina). Amazing photographs of these nebular objects have been obtained with the Hubble Space Telescope, large ground-based telescopes, and even amateur telescopes.

While HII regions trace the nebular environments of newborn star clusters, planetary nebulae mark the dying gasps of fully evolved stars. Stars with masses between 0.8 and 8.0 Suns evolve away from their core hydrogen-burning main-sequence states to become giants that are powered by the fusion of both hydrogen and helium deep within the stars. Internal instabilities ensue that end up driving the outermost layers to dissipate in the form of torrential winds. Eventually, the hot cores are exposed. Their ultraviolet radiation photo-ionizes the fleeing shells of dissipated gases, creating hot plasmas which then fluoresce with colors characteristic of the constituent ions. First discovered in the 1700s, these small emission nebulae were originally associated with planets – hence the name planetary nebulae. We now recognize them as being some of the most beautiful objects to grace the telescopic sky.

The third key type of emission nebula is the supernova remnant that marks the explosive demise of a once-massive star (or of a white dwarf). Stars with masses that exceed 8.0 Suns do not stop at fusing helium into carbon. They can go on to fuse carbon into oxygen, oxygen into silicon, and silicon into iron. Once iron is forged in the star's core, however, the game is up. No other heavier elements can be made without sucking energy out of the star. The dormant core rapidly collapses to become a neutron star or black hole, and the resulting release of gravitational energy drives a titanic explosion that completely expels the star's remaining layers. The nebular results are known as Type-II supernova remnants. For the first few thousand years, the shock-heated ejecta glow at all wavelengths – even at the highest-energy X-ray and gamma-ray extrema of the electromagnetic spectrum. The Crab nebula (M1) in Taurus is the most famous exemplar of this powerful early phase. Later on, the outward-rushing shock waves pile up and excite gases belonging to the ambient interstellar medium. The Veil nebula in northern Cygnus and the supernova remnant in the southern constellation of Vela are well-known versions of this latter phase.

Another type of supernova can occur when a white dwarf is in a close binary star system. At some point in the system's evolution, the white dwarf may feed on material from the close companion until it exceeds a critical mass and implodes. The release of gravitational energy will then ripple through the white dwarf, exploding it entirely. This Type-Ia supernova produces supernova remnants that are rich in heavy elements but lacking in hydrogen, the star's hydrogen layers having been dissipated long ago. The Tycho and Kepler supernova remnants are well-known examples of this genre.

Beyond the disk and into the halo, astronomers have found strange clouds of cold neutral atomic hydrogen slowly coursing through the Galaxy's outer reaches. These so-called "high-velocity clouds" (HVCs) are actually going around the Galaxy at much slower rates than the disk in which we reside. They only appear to be high velocity when seen from our swirling vantage point. We now know of about twenty HVCs and another ten or so intermediate-velocity clouds (IVCs) that appear to hew closer to the disk (see figure 7.8).

Although astronomers are not sure of their origins, HVCs appear to be mostly primordial in their elemental abundances and hence are likely vestiges of the original galaxy-building process. The big exception is the Magellanic Stream that spans much of the southern hemisphere. This HVC complex was likely stripped from the Large and Small Magellanic Clouds during close encounters between these galaxies and the Milky Way. Its material has intermediate elemental abundances, akin to those in the Magellanic Clouds themselves. Unlike most of the HVCs, the intermediate-velocity clouds may represent more chemically evolved material that was once part of the disk. Some unknown time ago, an intense spate of starburst activity in the disk expelled this material into the halo. The gas

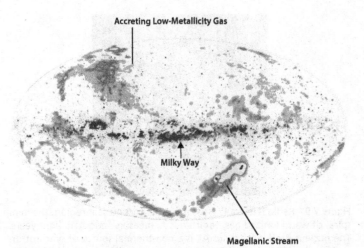

Figure 7.8 The Galactic halo contains enormous clouds of atomic hydrogen, as traced by radio astronomers. The Magellanic Stream is one of the largest, extending from the Large and Small Magellanic Clouds (companion galaxies to the Milky Way) over a full quadrant of the southern sky. (Adapted from image composite by Ingrid Kallick of Possible Designs, Madison, Wisconsin, with data from Bart Wakker, University of Wisconsin.)

is now falling back to the disk in the form of a vast Galactic "fountain."

Our Galaxy's dark heart

Until the development of radio interferometry in the 1960s and 1970s, very little was known about the center of our Galaxy. That is because the Galactic center is visibly obscured by a succession of dusty molecular clouds along our line of sight in the disk. Indeed, only about one visible photon in a trillion coming from the Galactic nucleus makes it past this gauntlet of gas

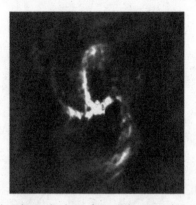

Figure 7.9 Radio telescopic view of our Galaxy's nuclear region. The mini spiral of warm ionized gas, Sagittarius A, measures about 15 light years. The nucleus itself, Sagittarius A*, is a non-thermal source of synchrotron radiation resulting from electrons moving at relativistic speeds in the presence of strong magnetic fields. It is barely resolved but can be no larger than the inner Solar System. (Courtesy of F. Yusef-Zadeh, D. A. Roberts and W. M. Goss, National Radio Astronomy Observatory, Associated Universities Incorporated, and the National Science Foundation.)

and dust to our telescopes. Radio emission, though, operates at much longer wavelengths and so can pass through the disk's dusty clouds without hindrance. By coordinating multiple radio telescopes into arrays that together span large areas, astronomers were able to achieve the angular resolution that is necessary to make revealing maps of the Galactic center. What they found is a complex of gaseous features that cannot be seen anywhere else in the Galaxy.

The excitement begins on an angular scale of about two arcminutes – equivalent to our Earthbound view of the largest craters on the Moon. At an estimated distance of 27,000 light years, this angular extent translates to a linear size of about 15 light years. Inside a fragmented ring of dense molecular gas, three arms of ionized gas appear to comprise a spiral "whirligig" that is centered on the Galactic nucleus (see figure 7.9). Infrared mapping of the hydrogen line emission from these arms has revealed motions consistent with the gas both rotating around and streaming towards the nucleus. The nucleus itself emits light in the form of synchrotron radiation – akin to that produced by our most powerful particle accelerators – where electrons whiz around magnetic field lines at near light speed. Ever more capable arrays of radio telescopes have shown the emitting nucleus to be no larger than 1/10,000 of a light year – the equivalent dimension of our Solar System's asteroid belt. We are thus dealing with an amazingly compact "machine" that can accelerate subatomic particles to relativistic speeds and so create powerful synchrotron radiation.

On a scale of 1/10 light year, infrared observations have enabled astronomers to track individual stars as they orbit the unresolved nucleus. Something interior to these orbits is driving the stars to whip around a common center with periods of only a few Earth years. By using simple Newtonian gravity, astronomers estimate that the mysterious gravitating agent has a mass equivalent to about 4 million Suns – all concentrated within a perimeter no larger than the orbit of Pluto around the Sun. Most astronomers are confident that we are seeing the dynamical effects of a supermassive black hole lurking in the center of our Galaxy. Other observed effects include strangely variable X-ray and gamma-ray emissions from the nuclear region. Though decidedly dim compared to the nuclear activity that has been observed in some other giant galaxies, our Galaxy's nucleus bears all the hallmarks of having been very active in the past. It may, in fact, have some surprises for us in the not-too-distant future.

8

Menageries of galaxies and their cosmic expansion

The history of astronomy is a history of receding horizons.
Edwin Powell Hubble, *The Realm of the Nebulae*

Like gigantic ships glowing in the night, galaxies give light, form, and substance to the looming darkness. Why nature has chosen these self-gravitating vessels to house the majority of luminous matter remains a mystery. One could imagine myriad stars and star clusters sprayed across the cosmos without any coordinated form or function – or much larger mega-galaxies spanning millions of light years and containing hundreds of trillions of Suns' worth of mass. Instead, astronomers have found that most observed galaxies have sizes spanning a few thousand light years to a few hundred thousand light years, with masses ranging from a few million Suns to a few trillion Suns. These relatively limited ranges of size and mass echo those of individual stars, where stellar sizes run from a tenth that of the Sun to a thousand times the Sun's radius, while their masses span a tenth that of the Sun to a hundred solar equivalents. What, then, delineates a galaxy? Issues of stability probably constrain the maximum sizes and masses of both stars and galaxies. For stars, the luminous output of these giant thermonuclear engines becomes the destabilizing factor. Beyond a hundred solar masses or so, the torrent of photons begins to overwhelm the star's gravity and so propels its gases to escape the star altogether. For galaxies, gravitational and tidal instabilities likely limit their maximum girths. At the opposite extreme, the minimum size and mass of a star is limited by its ability to fuse hydrogen in its core. Below about one-tenth of a solar mass, one is dealing with deuterium-burning brown dwarfs and non-fusing planets of even lower mass. By contrast, the minimal size and mass of a galaxy is not limited by concerns of powering. Instead, it was probably set by conditions shortly after the hot big bang, when most galaxies took form. We will consider these primordial times in chapters 9 and 10. But first, let's get better acquainted with the current-epoch galaxies that we can observe in detail.

Galaxies of the Local Group

The overwhelming majority of objects in the naked-eye sky belong to our Milky Way Galaxy. Every visible object in the Solar System, every star, star cluster, dark cloud, and emission nebula, shares our Galactic home in swirling co-existence. The rare exceptions to this home-grown pedigree can be counted on one hand. They include the Large and Small Magellanic Clouds, which are visible from the southern hemisphere; the Andromeda Galaxy (M31), which can be seen from the northern hemisphere on dark, moonless nights; and Andromeda's much fainter neighbor, the Pinwheel Galaxy in Triangulum (M33) (see figure 8.1). If you search online for "naked-eye galaxies," you will see a few more galaxies listed – but these are the province of very experienced naked-eye observers. By far the most prominent of the extragalactic objects in the visible sky are the Large Magellanic Cloud (LMC) and Small Magellanic Cloud (SMC), as they are located much closer to us than M31 or M33. Accounting for distance reveals that M31 closely matches the Milky Way in girth, followed by M33, the LMC, and SMC. The remaining forty or more galaxies in the Local Group are much smaller dwarf galaxies.

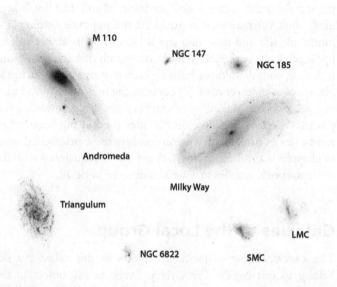

Figure 8.1 Relative sizes of the Milky Way, Magellanic Clouds, Andromeda Galaxy (M31), and Triangulum Galaxy (M33) – the largest galaxies in the Local Group – along with other notable dwarf galaxies. (Adapted from multiple sources.)

Galaxies beyond the Local Group

Our Local Group of galaxies represents just one of at least 100 galaxy groups that make up the Virgo Supercluster of galaxies (see chapter 3). Most of the galaxies that populate the well-known galaxy catalogues and atlases can be found in this larger structure. Distances to these galaxies are fairly well established, as Cepheid variable stars have been observed in most of them. And with the distances known, the sizes and luminosities of these galaxies have been determined with some confidence.

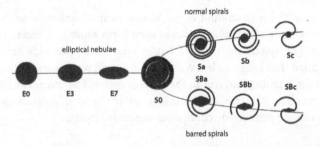

Figure 8.2 Hubble's tuning fork diagram for discriminating among galaxies according to their appearance in long-exposure photographic images. The "irregular" classification has since been added on the right-hand side between the forks. (Courtesy of Wikimedia Commons.)

Hubble's classification system

Initially, all that was known about these galaxies was their appearances as recorded on photographic plates. From deep images that he had taken with the 100-inch Hooker Telescope atop Mount Wilson Observatory near Los Angeles, Edwin Hubble was the first to make some sense out of the widely varying shapes. His classification system, first developed in 1926, led to the eponymous "tuning fork" diagram shown in figure 8.2.

From left to right in the diagram, the prominence of the central spheroid decreases, while that of the disk increases. We now understand that stars in the spheroid swarm around the galaxy's center, while those in the disk have mostly circular co-planar orbits. In the disk, the spiral windings are tightest for the Sa types and loosest for the Sc types. The two forks account for the unbarred and barred spiral types. The Milky Way Galaxy is thought to be an intermediate-type (bc) barred (B) spiral (S) galaxy with a Hubble classification of SBbc. Since Hubble's pioneering categorizations, astronomers have learned that the Hubble types follow a sequence of ever-increasing conservation of gas. The ellipticals contain the least amount of gas, having formed almost all of their stars within a few billion years of their birth some 12 billion years ago. The spirals are a mixed bag – with their spheroids having formed stars long ago but with their disks still retaining significant gas for ongoing starbirth in the current epoch. The irregulars have as much as 30 percent of their observable matter in gaseous form and so are primed for continuing starbirth activity.

Giant spiral galaxies

Of all the types of galaxies, the giant spirals reign supreme in their inspirational beauty. Wondrous color images by the Hubble Space Telescope reveal these galaxies to contain rich ecosystems of hot blue stars, glowing gaseous nebulae, and obscuring clouds of dust – with the most active regions arrayed along spiral arms like beads on a string or blossoms on a branch. Recently, amateur astronomers have made use of digital imaging technologies to make their own spectacular color images of these galaxies. Just consider the common names for some of the most famous giant spirals; the Pinwheel, Whirlpool, Sunflower, Black Eye, and Sombrero Galaxies have names that mirror our reactions to their exquisite structures.

So, how did these galaxies develop spiral structure? And how have they sustained their pleasing forms over cosmic time? Astronomers initially looked to the disks of these galaxies, where the spiral arms are located. What they found were stars and gas in shearing motion, such that the material in the inner disk of a spiral galaxy wheels around faster than the material occupying the galaxy's outer disk. The degree of shearing amounts to roughly 10 km/s per 1,000 light years of galactic radius – a small but significant effect compared to the characteristic rotation speeds of 100–300 km/s. This sort of shearing naturally produces spiral patterns in the disk. Just imagine a big cloud of star-forming gas.

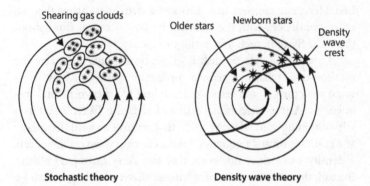

Figure 8.3 Comparison of the fragmentary spiral structure produced by the shearing of star-forming clouds of gas (left) and the more continuous spiral structure produced via the density wave theory as clouds of gas pass through the crest of a spiral density wave that has been established in the disk (right). (Adapted from *Galaxies and the Cosmic Frontier* by W. H. Waller and P. W. Hodge.)

The cloud – and newborn stars therein – will get shorn into a spiral fragment, with the outer stars lagging behind the inner stars. Put together a bunch of these star-forming clouds, and you can get a spiral-like pattern in the disk (see figure 8.3).

This sort of spiral structuring based on the shearing of star-forming clouds can explain raggedy spiral galaxies like the Pinwheel (M33) in Triangulum and the Sunflower/(M63) in Canes Venatici, where the spiral arms are short and fragmented. However, it cannot explain the "grand design" spiral galaxies such as the Whirlpool (M51), also in Canes Venatici, or the other Pinwheel (M101) in Ursa Major, where the spiral arms are both extensive and well delineated. Nor can it explain the striking rings of star-forming activity that have been found in some barred spiral galaxies such as M94 in Canes Venatici and M95 in

Leo. Here, astronomers have invoked a dynamical agent that has correspondences with the resonant behaviors extant throughout the Solar System, such as the rings of Saturn.

If a shearing galactic disk is gravitationally perturbed by the rotation of a stellar bar, or by a companion galaxy, or by its own spiral structure, it will respond according to its own natural frequencies. Astrophysicists have shown that the likely result is a self-sustaining spiral density wave that courses around the galaxy at a constant orbital frequency. Like a cresting wave on the ocean, a density wave does not consist of the same matter over time. Instead, the wave consists of whatever material it happens to be passing through.

For most disk galaxies, the stars and gas travel faster than the wave, and so they end up passing through the wave. The interaction is like a slow-traveling roadblock on a highway, where passing cars must slow down and consequently bunch up as they get past the blockage. Similarly, as stars and gas pass through the gravitating density wave, they slow down and concentrate. The gas clouds, in particular, aggregate and coalesce along the density wave crests to such a degree that they are primed to spawn new generations of stars. That is why we observe so many colorful collections of blue star clusters and roseate HII regions along the spiral arms.

The theory of spiral density waves in shearing galactic disks handily explains the observed substructure of spiral arms, where one can often see a spatial sequencing of pregnant dark dust clouds, newly formed crimson HII regions, and somewhat older blue stellar associations across the arm (see figure 8.3). Density wave theory also explains the star-forming rings that have been found in some spiral galaxies. These rings tend to occur at radii near resonances between the orbiting density wave and the orbiting gas clouds. Like Saturn's ring system, the orbital resonances clear out annular gaps and establish concentrations of matter nearby.

What remains unclear is the evolution of spiral galaxies over cosmic time. Do the star-forming spiral arms continue to slowly revolve around the galaxy according to the density waves that have been established in their disks? Or do the waves evolve and so transform the faces of these galaxies? Do the density waves themselves sap kinetic energy from the disk and so induce radial inflows of matter over billions of years? And what about the barred spirals? Are the bars relatively permanent pileups of stars in the responsive disk, or do they come and go? These sorts of questions continue to puzzle galactic astronomers and astrophysicists.

Giant elliptical galaxies

The biggest galaxies in the visible Universe are the giant and supergiant ellipticals. With sizes of several hundred thousand light years and luminosities of up to a trillion Suns, the giant ellipticals dominate the galaxy clusters in which they are typically found. At first blush, they appear rather simple in form and function. Just look at the giant ellipticals M84 and M87 that inhabit the core of the Virgo Cluster, and you will see very round, smooth, and yellowish stellar systems, with very little in the way of dust lanes or other features. The intensity of starlight falls off steeply with radius in a well-understood manner that can be modeled as a self-gravitating "gas" of stars that is swarming at a character-istic "temperature" around a very dense center. Occupying that center, in each case, is a supermassive black hole. For the giant and supergiant ellipticals, the mass of that black hole can range from millions to billions of Suns. So begins our introduction to the odder aspects of these seemingly benign galaxies.

Another clue to their strangeness is immediately apparent upon closer inspection. Images taken by the HST and other world-class telescopes have revealed powerful jets of gas that are shooting from the centers of some giant elliptical galaxies. A good case in point is the supergiant galaxy M87 in Virgo. Its jet can be followed optically for 1,500 light years and at radio wavelengths for an incredible 250,000 light years. Breaks in the jet indicate episodes of outburst activity followed by relatively quiescent periods. All this eruptive activity can be traced to the galaxy's nucleus, where a 5 billion solar-mass black hole is thought to reside.

The final major clue to the nature of giant elliptical galaxies has come from very long-exposure images, wherein the faint outer parts of these galaxies can be revealed. Beginning in the 1980s with deep photographs and continuing with even deeper digital images, astronomers have found concentric shells of diffusely distributed starlight. Moreover, the shell pattern on one side of the galaxy is often observed to be interleaved with the shell pattern on the opposite side. Astrophysicists have successfully modeled these faint shell patterns as tracing the trajectories of galaxies that have been captured and consumed by the giant elliptical galaxy. As the victim galaxies spiraled into the predatory giant, they left behind shell-like vestiges of themselves wherever they reached the extremities of their eccentric elliptical orbits. The interleaving of stellar shells is consistent with the inward spiraling of these hapless galaxies.

Starburst galaxies

While the giant elliptical galaxies are thought to have cannibalized much smaller galaxies, starburst galaxies seem to occur most often when two galaxies of similar girth interact. Another important difference is that at least one of the interacting galaxies must have conserved robust quantities of gas. It is the gas, once shepherded into dense clouds, that produces the riots of newborn stars and their energetic consequences so characteristic of starburst activity. Consider the Cigar Galaxy (M82), the nearest galaxy to us that is currently experiencing a powerful starburst in its dense center. Gravitationally provoked by its much bigger neighbor Bode's Galaxy (M81), M82 is afire with massive clusters of hot blue stars, myriad radio supernova remnants, and a huge bipolar outflow of ionized gas. It is currently forging new stars at an unsustainable rate. Either it will quiet down very soon, or it will run out of star-forming gas in less than a few hundred million years.

Other starbursting galaxies are more closely paired and in the process of merging (see figure 8.4). Astronomers believe that these provide precious prototypes of the galaxies that may have characterized the more crowded early Universe, when the giant elliptical galaxies and the bulges of what would become giant spiral galaxies were just beginning to take form.

Galaxies with active nuclei

We end this chapter with the most bizarre objects in the visible Universe, the active galactic nuclei that comprise roughly one percent of all giant galaxies at the current epoch. Beginning in the 1950s with the discovery of strangely luminous radio sources in the sky, astronomers began to match these brilliant radio-wave emitters with visible galaxies and so identify a menagerie of aberrantly active galaxies. As the angular resolution and sensitivity of radio telescopes improved, the resulting maps revealed enormous bipolar jets of emitting gas extending for upwards of several hundred thousand light years. Often, the optical galaxy in the center would be dwarfed by these amazing outflows by factors of ten or more. Many of these galaxies turned out to be giant ellipticals, or strangely distorted mergers of previously intact galaxies.

Meanwhile, astronomers were finding that some apparently normal spiral galaxies hosted brilliant nuclei which radiated

Figure 8.4 Schematic of two galaxies in the process of merging. Many starbursting galaxies occur in such closely interacting systems. (Courtesy of F. Zwicky in *Physics Today*, vol. 6 [1953], p. 7.)

mostly at optical and infrared wavelengths. These were dubbed Seyfert galaxies after Carl Seyfert, who first described this class in 1943. The spectra of these sources displayed broad emission lines of highly ionized atoms. The high ionization state indicated gas clouds in the presence of some incredibly hot energizer. The broad profiles of the spectral emission lines further indicated extreme Doppler shifting in wavelength of the emission from the gas clouds, due to the high velocities of these clouds. Finally, the emission was seen to fluctuate in luminosity on timescales of a few hours to a few days.

Piecing it all together, astronomers have since concluded that these observations are collectively consistent with the Seyfert galaxies hosting supermassive black holes that are surrounded by disks of hot accreting matter. Each accretion disk is like "ground zero," where infalling gas clouds collide with other matter in the disk. The resulting shock-heated maelstrom glows brilliantly at all wavelengths, thus ionizing and energizing any gas clouds that surround the disk. Variations in the accretion rate explain the observed fluctuations in luminosity from both the accretion disk and the responding gas clouds.

Sometimes, visible light from the brilliant accretion disk is not evident. Instead, strong infrared radiation is observed. Astronomers call these sorts of systems Seyfert 2 galaxies and model their nuclei as being surrounded by thick rings of obscuring dust. The dust absorbs the light from the accretion disk – thus blocking our visible view of it – and re-emits it at mid-infrared wavelengths. The best nearby example of such a galaxy is the barred spiral M77 (NGC 1068) in Cetus the whale, whose central region copiously emits in the infrared, and whose spectral line emission is thought to arise from highly ionized gas clouds located no more than a light year from the nuclear black hole and accretion disk.

If our view of the nuclear region is more face-on than edge-on, we get to see the accretion disk's direct emission. Here we are witness to the highest-energy radiation and the broadest spectral emission lines. Systems of this ilk are called Seyfert 1 galaxies – and they serve as relatively nearby proxies for the even more luminous quasars (or QSOs) that are evident at much greater distances.

We have to go pretty far afield to find a bona fide quasar, an active galaxy whose nuclear emission overwhelms all other emissions. One of the nearest examples – and the first to be identified as a quasar – is 3C 273 in the constellation of Virgo. The high redshift of its spectral emission lines indicates that the cosmos has expanded by a factor of 1.16 since its light was first emitted. That means we are seeing it as it was emitting roughly 2.4 billion years ago, when it was about 2–3 billion light years away from us. Though incredibly remote, 3C 273 can still be observed by amateur astronomers with telescopes that have apertures of fourteen-inch diameter or more. Its apparent magnitude of 12.9 combined with its great distance yields an absolute luminosity that is simply nuts – about 4 trillion Suns, or more than 100 times that of a normal galaxy, all from a region no larger than our Solar System. Recent imaging from the HST and Chandra X-ray Observatory has provided hints of structure from 3C 273, including a powerful jet and some surrounding fuzz of starlight. Images of other quasars show a wide variety of shapes, from seemingly normal spirals to multiple merging galaxies. We will revisit quasars in chapter 9, as they appear to occupy a special period of time in the unfolding Universe.

REDSHIFT

The redshift of a spectral emission line from a galaxy is defined as ($z = \Delta\lambda / \lambda_o$), where $\Delta\lambda$ is the observed increase in the emission line's wavelength, and λ_o is the wavelength as measured in the laboratory. This quantity traces the amount the Universe has expanded since the light was first emitted. In the specific case of the quasar 3C 273, the observed redshift is $z = 0.158$. That means the Universe has expanded by 16 percent, or by a factor of 1.16, since the observed light was first emitted. Assuming the Universe to be 13.8 billion years old (and essentially undergoing free expansion), we are seeing the light as it was emitted about 2.4 billion years ago.

Counting galaxies

As this chapter outlines, the observable Universe abounds with a wondrous variety of galaxies. Astronomers are fueled by that wonder and – through careful multi-wavelength imaging and spectroscopic studies – have revealed a complex interplay between the stars, gas, and active nuclei within these realms. Moreover, they have begun to piece together the sundry evolutionary tales being told by these "island universes" over cosmic time (see chapter 10). But there is so much more to learn first, beginning by simply counting the number of galaxies that we can observe. This was first successfully accomplished in 1996, after the HST had imaged the same piece of sky for a record-breaking ten days of continuous exposure. The resulting "Hubble Deep Field" captured only a tiny mote of sky situated in the constellation of Ursa Major. Imagine viewing a dime held at arm's length and fixating on Franklin Roosevelt's eye; that would be tantamount to the small field of view the HST imaged. However, in that image were thousands upon thousands of galaxies. This incredible "core sample" of the galaxian Universe opened up all kinds of research opportunities – including the prospect for estimating the total number of galaxies that inhabit the visible cosmos.

Suffice to say, the number of galaxies within our observational reach is truly astounding – from 50 to 100 *billion* of them. And we are only counting the giant galaxies and the starbursting dwarfs here. The smaller and less active galaxies will miss our probing. Since 1996, other deep fields have been imaged by the HST, Spitzer Space Telescope, and several ground-based telescopes. We now have a multi-wavelength dossier of galactic core samples that can be accessed to explore the evolution of galaxies over cosmic time – from more than 10 billion years ago to the present day. To actually detect "first light" from the youngest galaxies, however, we will have to wait for the James Webb Space Telescope and other telescopes that have been designed to harvest this faint primordial radiation (see chapter 10).

The cosmic expansion

Well before the Hubble Space Telescope and its stunningly deep fields, astronomers in the 1920s had begun scrutinizing the galaxies that were available to their telescopes and detectors. By breaking up the galactic light into spectra and photographing these spectra with long exposures, they discovered that the spectral lines showed significant Doppler shifts in wavelength. The blueshifted spectra indicated motions towards us, while the redshifted spectra indicated receding motions. For example, the Andromeda Galaxy (M31) showed a spectral blueshift caused by the Andromeda and the Milky Way Galaxies approaching each other at a velocity of 110 km/s. Similarly, the Triangulum Galaxy (M33) also sports a spectral blueshift which has it approaching us at a speed of 44 km/s.

Beyond the Local Group of galaxies, however, most of the galaxy spectra showed redshifts. Edwin Hubble's able assistant Milton Humason was especially adept at recording these spectra on photographic plates using the aforementioned Hooker Telescope in southern California – the largest telescope in the world at the time. Working with these spectra and relating their redshifts to the best estimates of how far away the corresponding galaxies were, Hubble obtained his seminal relation in 1929. It states that the greater the galaxy's distance, the greater its corresponding redshift. Hubble interpreted this redshift as a recessional velocity, in keeping with the optical Doppler effect. This was sufficient for him to make the great conceptual leap to interpret his law as evidence for an expanding universe of galaxies. Here, illustrative analogies help to vivify his reasoning.

Consider a loaf of raisin bread before and after it is baked. With respect to one randomly chosen raisin, every other raisin is at a particular distance from it. Some raisins are very close to the reference raisin, while others reside at greater distances towards the opposite side of the loaf. Once the loaf is baked, it has expanded to a greater volume. The closest raisins have moved away from the reference raisin by only a small amount, but those raisins that were initially more distant will have moved away by a greater amount, and the few raisins at the far perimeter of the loaf will have moved away by the greatest amount. This distance–motion relationship works for any randomly chosen reference raisin. In other words, it is indicative of a universal expansion.

Today, astronomers no longer interpret the galaxy redshifts as arising from a kinematic model, whereby the galaxies move away from us at velocities that trend with the respective distances. Instead, they see the stretching wavelengths of light as tracing the stretching of space itself. Rather than the galaxies traveling through space, it is the fabric of space that is expanding and taking the galaxies along for the ride! Fortuitously, the picture of an expanding loaf of raisin bread works even better as an analogy of this interpretation.

Correct interpretations notwithstanding, the rate of this expansion continues to be expressed in kinematic units of km/s/Mpc (Mpc being short for megaparsecs, where 1 Mpc = 1 million parsecs or 3.26 million l.y.). Our current best estimate for the

DIFFERING VALUES OF THE HUBBLE CONSTANT

Recent measurements from diverse approaches have each reduced the uncertainty in their respective values of the Hubble constant. However, the values themselves now differ – perhaps significantly. As I write this paragraph, the use of Type Ia supernovae as standard candles for measuring galaxy distances yields a value for H_o of 74.0 ± 1.5 km/s/Mpc. However, details of the cosmic microwave background indicate a value of 67.5 ± 0.5 km/s/Mpc, while a new distance-measuring technique involving the gravitational lensing of distant quasars by more nearby galaxies has produced a range of values from 72.5 ± 2.2 km/s to 82 ± 8 km/s/Mpc. Scientists don't yet know whether the seeming disparities call for some sort of new physics, or whether systemic errors and internal biases can explain the conflicting results.

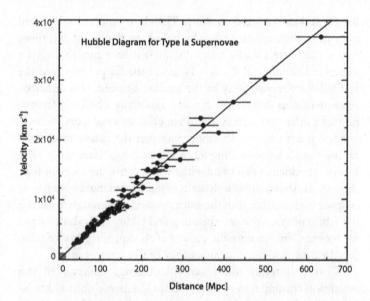

Figure 8.5 A recent plotting of the Hubble Relation between a galaxy's distance in megaparsecs (Mpc) and its recessional velocity (km/s). The slope of the relation gives the value of the Hubble constant. The square in the bottom left corner of this diagram denotes the space that Edwin Hubble sampled in his pioneering study. (Adapted from R. P. Kirshner, *PNAS*, vol. 101 [2003], pp. 83–13.)

so-called Hubble constant (H_0), using the most precise measurements of galaxy distances, is:

$$H_0 = 72 \pm 6 \text{ km/s/Mpc}$$

while, according to the kinematic interpretation, the recessional velocity (v_r) is related linearly to the distance by $v_r = H_0 \times d$, as plotted in figure 8.5.

Hubble's work continues to be invaluable to the field of astronomy. His eponymous relation establishes the existence of an expanding Universe, but also can be used to derive distances to remote galaxies, galaxy clusters, and superclusters, as we saw in chapter 3. Further scrutiny of it yields a possible non-linearity, which has been used to claim the Universe's rate of expansion is accelerating; as we will see in chapter 13, such an accelerating expansion would lend further credence to the presence of some sort of dark energy permeating all of space. Before that, however, we go back to where it all began, and there we will see that the Hubble constant can be used to obtain the Hubble Time, a first approximation for the age of the Universe.

OUR MOMENT
IN TIME

9

The hot big bang

In the beginning was the Creation of the Universe. This has made a lot of people angry, and has been widely regarded as a bad move.
Douglas Adams, *The Hitchhiker's Guide to the Galaxy*

Once upon a spacetime, our Universe emerged from the quantum vacuum. This vast, dark ocean of possibilities may have spawned other spacetimes – so physically isolated from each other that we may never ever detect them. Nonetheless, theoretical cosmologists have speculated upon such a multiverse and how it might be configured. Is the multiverse infinite, so that every realization in our particular Universe – including what you were thinking about over breakfast – is being repeated in other spacetimes ad infinitum? Or is the multiverse constrained to specific sets of dimensions, physical constants, and laws? We may never know, but stables of cosmic theorists continue to sweat the odds.

Setting the stage

Meanwhile, back in our evolving spacetime, we have our own questions. Is our Universe infinite in spatial extent, or does it have an actual size that has grown from a submicroscopic seed some 13.8 billion years ago to what we see today? That our Universe has a finite age sets a limit on how much of it we can actually perceive. Anything exceeding this limiting radius in spacetime has not had sufficient time for its light to reach us. Therefore, our limiting horizon (usually expressed as our limiting lookback time) is thought to have a radius equivalent to 13.8 billion years of light travel. It is because we can see nothing beyond this limiting radius that our night sky is essentially dark. Consider the alternative. Were the Universe infinite in both age and extent, every sightline would ultimately intersect the surface of a star – thus producing a brilliant sky at all times. This conundrum, known as Olbers' Paradox, puzzled astronomers until they allowed for a Universe of finite age.

Beyond the Andromeda Galaxy and the Local Group of galaxies, universal expansion complicates the metric distances.

WHAT IS LOOKBACK TIME?

Because light travels at a finite speed of 300 million meters per second (300,000 kilometers per second, or 186,000 miles per second), it takes a finite time for light from a radiating source to reach us. That finite time is known as the source's lookback time. For example, we see the Moon (384,000 kilometers away) as it was reflecting the Sun's light 1.28 seconds ago; therefore, it has a lookback time of 1.28 seconds. We see the Sun itself (150 million kilometers away) as it was roiling and emitting 8.33 minutes ago, meaning that the Sun has a lookback time of 8.33 minutes. Moving out, here are some lookback times worth considering:

Saturn (1.1 hours at closest approach to Earth)
Pluto (6.9 hours at closest approach to Earth)
Alpha Centauri (4.2 years)
Vega (25 years)
Polaris (434 years)
Orion Nebula (1,500 years)
Galactic center (27,000 years)
Andromeda Galaxy (2.5 million years)
Nearest quasar – 3C 273 (2.4 billion years)
Farthest detectable galaxy – GN-z11 (13.4 billion years)

For example, the distance to a particular galaxy at the moment when the light was emitted ends up being much less than the distance at the moment of detection. That is why the lookback time is the favored means of quantifying cosmic distances, as it best incorporates the expansion and so gives us a single measure of the object's distance from us within the context of the expanding cosmos. Perhaps surprisingly, astronomers using the Hubble Space Telescope and the largest ground-based telescopes have detected galaxies whose light is so redshifted today that it was likely emitted when the visible Universe was more than ten times more compact and so was only a few hundred million years old. What we can see from those primeval epochs is in many ways dramatically different from what we observe in the current-epoch Universe, with the early-epoch galaxies appearing much smaller and lumpier than their contemporary counterparts.

In the subsequent chapters, we will stick with the part of our Universe which dwells within the limiting light-radius set by our Universe's finite age of 13.8 billion years, and is thus detectable to us. That limitation in spatial extent does not prevent us from exploring the full sweep of cosmic history, however. All the seminal stages, from the hot big bang to the epoch of recombination, to the formation of the first galaxies, stars, and planets, are observable – at least in principle.

Cosmic genesis

That our Universe had a beginning is about as uncanny a statement as one could ever proclaim. Yet that has not prevented people from crafting and communicating narratives of cosmic origin as essential parts of their cultural identities. Indeed, genesis stories abound across civilizations going back well before the advent of writing. Perhaps uniquely among terrestrial animals, we humans need to explain our own existence in terms of the greater Universe's origin and evolution.

I am old enough to remember when astronomers were still uncertain whether the Universe actually had a beginning. In the 1960s, two cosmological theories were vying for primacy. One was called the Steady State theory. Championed by Sir Fred Hoyle and colleagues in the 1940s and 1950s, it recognized the cosmic expansion that was discovered by Edwin Hubble in 1929, but allowed for new matter to fill in the voids. With this theory, the Universe remained essentially the same throughout space and time, thus adhering to the popular notion of uniformity – what was known as the perfect cosmological principle. There would be no special place or time, including our own time here on Earth.

The other theory, first popularized by George Gamow, also in the 1940s and 1950s, posited the opposite picture, whereby our Universe has been expanding, thinning out, and cooling following an epoch of incredibly high density and temperature several billions of years ago. This hot big bang theory also considered the expansion rate quantified by the Hubble constant (H_o) but inverted it, thus backtracking the expansion to its torrid beginnings. The so-called Hubble Time ($T = 1 / H_o$) that has since elapsed, using the best estimates of the Hubble constant then available, ranged between 10 and 20 billion years. The shorter timescale had its difficulties, as ages of some globular star clusters appeared to exceed this hard limit. How could a star cluster be older than the Universe that spawned it? The two theories continued to draw their respective adherents until 1964, when the cosmic microwave background (CMB) radiation was discovered.

Here was the "smoking gun" that showed the primeval Universe to have been radically different from what we see today. As early as the 1940s, physicists had predicted this remnant radiation from the Universe's hot beginnings. They further recognized that the wavelengths of the radiation would have been stretched out by the universal expansion. What began as an orange glow from an ionized plasma at a temperature of several thousand degrees would be transformed into a faint microwave "hiss" at an equivalent temperature of just a few degrees above absolute zero. That is what Arno Penzias and Robert Wilson detected while working with a fifty-foot-long horn antenna situated atop Crawford Hill in Holmdel, New Jersey. As scientists with Bell Telephone Laboratories, Penzias and Wilson were testing the antenna for the purposes of satellite communications and doing some radio astronomy. At first, they thought the signal was noise coming from some part of the antenna. Only after eliminating all possible sources of noise, including bird droppings inside the horn antenna, did they conclude that the signal was cosmic. This all-sky "background" emission has since been recognized as the remnant glow from the hot big bang, an amazing discovery that netted Penzias and Wilson the 1978 Nobel Prize in Physics.

Since its discovery, the CMB has been measured and characterized with ever-increasing acuity. Its spectrum follows that of an ideal thermal radiator (black body) at an equivalent temperature of 2.725 kelvin (see figure 9.1). Astrophysicists regard this radiation as being most consistent with a Universe that has since expanded by a factor of 1,100 over a time period of 13.8 billion years. What we are detecting is radiation from matter that has just cooled from an ionized plasma state to a neutral atomic state. At that critical epoch of recombination, a mere 380,000 years after the hot big bang, the Universe became transparent to its own radiation, and so the photons of light were set free to traverse the expanse and be detected by our instruments. That the current best estimate of the Hubble constant (72 km/s/Mpc), the corresponding Hubble Time (13.6 Gyr), and the appropriately younger age of the oldest globular clusters (12.7 Gyr) jibe so well with the CMB-derived age of 13.8 Gyr for the Universe continues to encourage astrophysicists that they are on the right track.

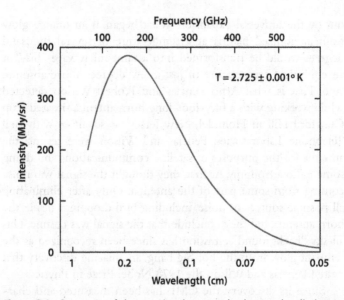

Figure 9.1 Spectrum of the cosmic microwave background radiation, as measured by the Cosmic Background Explorer (COBE) in the early 1990s. This spectrum is identical to that of a perfect black body at a single temperature of 2.725 kelvin. (Courtesy of C. Bennett, DMR, COBE, GSFC, NASA.)

In order to obtain a true picture of the CMB, astrophysicists have had to carefully subtract the "foregrounds" of microwave emission from our Milky Way Galaxy and from the myriad other galaxies in the sky. They also have had to compensate for the Doppler shifting of the black body radiation across the sky due to the Solar System's motion around the Milky Way and the Milky Way's motion with respect to the cosmic reference frame. It turns out that the remaining CMB is incredibly smooth – with a degree of fluctuation that is about 1 part per 100,000. A freshly groomed skating rink comes close to this degree of smoothness. Most of the fluctuations occur on an angular scale of about a degree on the sky (see figure 9.2).

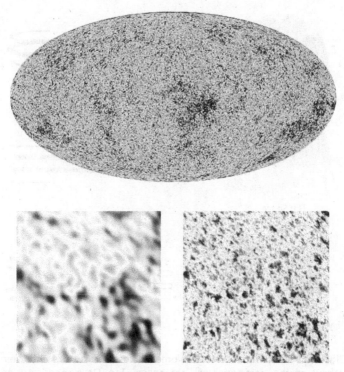

Figure 9.2 *Top:* All-sky mapping of the cosmic microwave background, as seen by the Wilkinson Microwave Anisotropy Probe (WMAP), after all Galactic and other "foregrounds" have been removed. *Bottom:* The left close-up is by the WMAP; the right mapping of the same region is by the more recent Planck satellite. Both maps show similar mottling on angular scales of a degree – consistent with the imprint of acoustic waves coursing through an ionized plasma just when the expanding cosmos cooled to a neutral atomic state. Overdensities in the plasma are evident as relatively darker regions. (Courtesy of ESA and the Planck Collaboration; NASA/WMAP Science Team.)

This characteristic angular spacing plus other less prominent peaks in the distribution of spacings tell astrophysicists that the Universe is exquisitely "flat." That means two laser beams shot

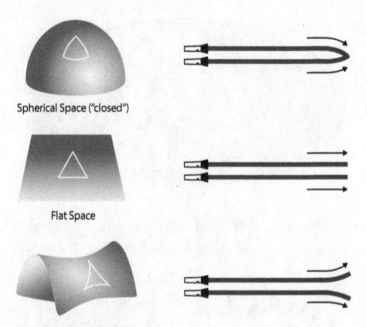

Figure 9.3 Two-dimensional analogies of cosmic curvature and their effects on the trajectories of light beams. *Top:* On a spherical – or "closed" – surface, two initially parallel lines will converge. Initially parallel laser beams would act similarly. *Middle:* On a flat surface, the parallel lines and laser beams will remain parallel into perpetuity. *Bottom:* On a flaring – or "open" – surface, the initially parallel lines and laser beams will diverge. (Adapted from *Discovering the Universe* by N. F. Comins and W. J. Kaufmann III, 4th edition, W. H. Freeman [1997].)

into the firmament parallel to one another will never converge nor diverge. Larger angular spacings in the CMB would indicate converging beams (like the longitude lines on a globe), while smaller spacings would indicate diverging beams (like lines drawn on a flaring saddle) (see figure 9.3). The near-perfect flatness of space, in turn, indicates that our Universe is dominated by some sort of dark energy. Otherwise, the estimated amounts of gravitating ordinary matter and dark matter in the Universe would be insufficient to flatten its form. Perhaps most exciting, the distribution of spacings bespeaks primeval conditions, when the Universe was just emerging from the quantum vacuum. Even then, the cosmos had to be exquisitely flat, the same in all directions (isotropic), and homogenous to a fault. These constraints present serious challenges to any cosmologist trying to figure out how it all started.

Key epochs

By considering the expanding Universe and running the clock back to the earliest times, we can imagine a cosmos that was far denser and hotter – what George Gamow in 1952 called the primeval fireball. Physicists have done exactly that and have thereby found critical epochs in the unfolding Universe of space, time, matter, radiation, and other, still-mysterious forms of matter–energy (see figure 9.4).

The Planck epoch

Time $\approx 10^{-43}$ seconds, temperature $\approx 10^{32}$ kelvin

Here be dragons. We don't really know much about this epoch, because space and time had yet to emerge as separate dimensions – the cosmic clock had yet to start ticking. Moreover, all of the known forces that are recognized today (the strong, weak, electromagnetic, and gravitational forces) were as one. Only after this epoch transpired could space, time, and gravity "freeze" out of the chaos and so set the stage for the Universe to unfold. Physicists imagine fundamental particles and antiparticles popping in and

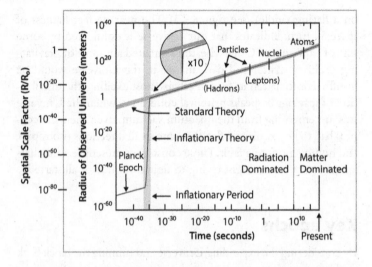

Figure 9.4 Schematic timeline of the primeval Universe, including key epochs in the expanding and cooling meta-system of matter–energy. (Adapted from A. Guth, "Inflation and the New Era of High-precision Cosmology," *MIT Physics Annual* [2002].)

out of existence amid the quantum "foam" of possibilities, but that's about it. Perhaps someday we will have a viable theory of quantum gravity. But until then, the Planck epoch will remain beyond our ken.

The inflationary epoch

Time $\approx 10^{-35}$ seconds, temperature $\approx 10^{27}$ kelvin

Once the Universe had expanded and cooled past the black hole of unknowing that we call the Planck epoch, a broth of fundamental particles inhabited the emergent Universe. These included quarks, gluons, leptons, their anti-particle counterparts, and photons – all in a stew of mutual interactions and transformations. A host of exotic particles including axions, magnetic monopoles, sterile neutrinos, and gravitons have been posited for this time as well, but none of these has yet been confirmed. Meanwhile the Universe had cooled sufficiently for the strong force to break away from the prior mash-up of strong, weak, and electromagnetic forces (the basic forces that respectively govern the internal behavior of particles, nuclei, and atoms). This liberation is thought to have driven a drastic inflation of the Universe, whereby it doubled in size more than 150 times in less than 10^{-33} seconds (see figure 9.4). What began as a sub-microscopic blip ended up as a macroscopic cosmos that was 50 orders of magnitude larger. (This grapefruit-sized realm has since expanded, at a far more leisurely rate, by another 30 orders of magnitude to become today's observable Universe.) The freezing out of the strong force also ramped up the production of elementary particles and photons – what gave the hot big bang its oomph.

Cosmologists favor this inflationary scenario, as it solves several problems with the hot big bang theory. Namely, it drastically inflates the observable Cosmos, so that whatever spatial curvature it had is rendered moot – thus solving the flatness problem. It also ensures that all directions in the observable cosmos share the same structuring – thus solving the isotropy and homogeneity problems. Moreover, it predicts that quantum fluctuations during this epoch could have grown into the acoustic record of fluctuating density and temperature that is evident in the cosmic microwave background. As of this writing, astronomers have sought other tell-tale signs in the CMB that would further confirm the inflationary epoch but have yet to get the clincher. If/when that happens, the theory's authors likely will merit a Nobel Prize.

The particle epoch

Time $\approx 10^{-12}$ seconds, temperature $\approx 10^{15}$ kelvin

Before this time, elementary particles and their antiparticles had been locked into an existential dance of creation and mutual annihilation. But once the Universe had expanded and cooled enough, stable particles could emerge. The first stage of this particle epoch was the hadron epoch, which refers to the binding of quarks into protons, neutrons, and mesons. Three quarks make up each proton and neutron, while two quarks comprise each meson – all held together by gluons. Simultaneously, the combined "electroweak" force bifurcated into the electromagnetic and weak forces, further enabling the production of W and Z bosons out of the chaos. A very slight excess of matter over antimatter (by 1 part in a billion) eventually led to the material Universe that we recognize today.

The stabilization of particles continued in the following lepton epoch (time $\approx 10^{-4}$ s, temperature $\approx 10^{11}$ K) when electrons, neutrinos, and other relatively light particles broke free from the morass. This important period came to an end at around one second after the big bang.

The nuclear epoch

Time $\approx 10^2$ seconds, temperature $\approx 10^9$ kelvin

Once the expanding Universe cooled to a temperature of a billion kelvin, its protons and neutrons could begin binding into various nuclei. These included the stable nuclei of deuterium (1 proton + 1 neutron), helium-3 (2 protons + 1 neutron), helium-4 (2 protons + 2 neutrons), and lithium-7 (3 protons + 4 neutrons). This period of nucleosynthesis lasted about 20 minutes, after which any free neutrons would have decayed back into protons and electrons, thus nixing further reactions. One of the greatest successes of the hot big bang theory is its accurate prediction of the relative abundances of these various atomic nuclei in the allotted time. Even before stars began to do their own element building, the primordial Universe was already endowed with 75 percent hydrogen-1 (whose nucleus comprises a single proton), about 25 percent helium-4, 0.01 percent deuterium and helium-3, and trace amounts (of the order of 10^{-10}) of lithium-7.

The matter epoch

Time $\approx 10^{11}$ seconds–10^4 years, temperature $\approx 10^5$ kelvin

Before this epoch, the energy budget of the Universe was dominated by photons of light. The incessant cosmic expansion ultimately changed this status quo, however. As the Universe swelled, the number density of both photons and material particles diluted in accordance with the increasing volume – and thus the cube of the increasing size (R^3). Meanwhile, the energy per photon declined as each photon's corresponding wavelength stretched with the expanding Universe, i.e. as the size (R). The combined spatial dilution of photons and energy reduction per photon meant that the energy density of photons waned as the fourth power of the increasing size (R^4) compared with the third-power (R^3) dilution of matter. It took about 10,000 years of cosmic expansion for the photon energy density to fall below that of the material particles. Today, we inherit this matter-dominated Universe.

The atomic epoch

Time $\approx 10^{13}$ seconds–4×10^5 years, temperature $\approx 3 \times 10^3$ kelvin

This epoch immediately follows the recombination of electrons and nuclei to form atoms with no net charge. Details of the cosmic microwave background radiation have enabled observational cosmologists to reckon the age of this epoch (380,000 years after the big bang), its temperature (3,000 kelvin), and the factor by which the Universe has expanded since this epoch (1,100). They interpret the faint all-sky microwave glow as the "surface of last scattering" from the primeval plasma just when it neutralized and became transparent to photons. A reasonable analogy is the visible surface of the Sun: we cannot see through the Sun into deeper layers of the solar plasma, but must content ourselves with seeing its surface of last scattering, where the density has dropped sufficiently for the photons to escape. Similarly, the CMB is thought to represent the critical moment when electrons could be bound to nuclei and so form neutral atoms. The photons permeating the cosmos no longer had a plasma to interact with and so could propagate through space and time to reach our detectors 13.8 billion years later.

Coda

The hot big bang theory accounts for an awful lot that astronomers have found out about the Universe, including:

1. The cosmic expansion. Rewinding this expansion toward the beginning leads to a cosmos that was much denser and hotter. Best estimates for the expansion age are around 13.8 billion years.
2. Olbers' Paradox. The night sky is dark because the Universe has a finite age. Light from anything beyond the lookback horizon of 13.8 billion light years has yet to reach us.
3. Cosmic abundances. The observed relative abundances of hydrogen, deuterium, helium-3, helium-4, and lithium-7 are predicted outcomes of the nucleosynthesis that transpired during the first twenty minutes of the hot big bang.

4. The cosmic microwave background radiation. This all-sky glow has been interpreted as the relic emission from the Universe when it was just cooling from an ionized plasma state to a neutral atomic state, some 380,000 years after the hot big bang.

By itself, the hot big bang theory cannot explain the exquisite flatness, isotropy, and homogeneity of the Universe that is inferred from details of the cosmic microwave background. That's where the inflationary epoch comes in. By drastically growing a mote of spacetime at the earliest stage of our Universe's emergence, cosmic inflation accomplishes the necessary smoothing of the matter and radiation that would evolve into the Universe we recognize today. Cosmic inflation has yet to be verified experimentally, but many astrophysicists believe that future measurements of polarization in the CMB radiation will soon seal the deal.

This chapter ends with the atomic epoch, when the Universe consisted of neutral atoms amid a flurry of visible photons at a temperature of about 3,000 K. I would be remiss to neglect the ever-mysterious dark matter and dark energy, however. Indeed, the formation of galaxies, galaxy clusters, and superclusters out of the minuscule overdensities that are manifest in the CMB appears to depend especially upon the disposition of the dark matter. In the next chapter on galaxy formation, the gravitating role of dark matter and several other seminal issues will be discussed.

10

The emergence of galaxies

In all chaos there is a cosmos, in all disorder a secret order.
Carl Jung, *Modern Man in Search of a Soul*

The challenge of galaxy formation can be found in the tiny fluctuations that are evident in the cosmic microwave background radiation. If interpreted as material over- and under-densities of only a few parts per 100,000, how could these hints of structure amplify into the much denser galaxies, galaxy clusters, and super-clusters that inhabit today's Universe? Moreover, the amplification had to happen rapidly. The Hubble Deep Field and more recent deep fields have revealed galaxies at redshifts of five or more, which indicate formation ages less than a billion years after the big bang. How could all this drastic condensation of matter have happened so quickly?

Unfortunately, the telescopes that would be powerful enough to probe the so-called Dark Ages between 400,000 and 1 billion years after the big bang have yet to be built. The Hubble Deep Fields have provided us with some glimpses, and these suggest the first galaxies were relatively small, oddly shaped, and under-going rampant starburst activity. This observational impasse will ease somewhat in the next decade, when the James Webb Space Telescope (observing in the infrared), the Square Kilometer Array

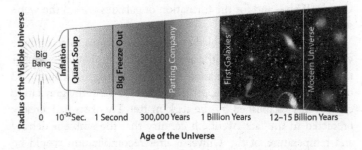

Figure 10.1 Graphical representation of the evolving Universe – including the Dark Age, epoch of galaxy formation, reionization, and subsequent transformations into the current epoch. (Courtesy of NASA).

(observing in the radio), and other pioneering new telescopes are fully deployed.

Simulated scenarios

Meanwhile, tremendous strides have been made in the last decade by teams of astrophysicists developing numerical simulations. Using the most powerful supercomputers, multiple groups of scientists have conceived varied recipes for building galaxies out of the expanding and cooling broth of dark matter and atoms (see figure 10.1). They typically begin with the dark matter, as it could begin to gravitationally coalesce even before the epoch of recombination. The ordinary matter was then still in an ionized plasma state and so was prevented from congealing by the torrents of photons that kept interacting with it. By contrast, the unaffected dark matter could go its own way and gravitationally respond to any primordial fluctuations rippling through the cosmos at this time. The dark matter's predicted condensation into enormous clumps and filaments would set the stage for the

Cosmic Web – and for the formation of galaxies within the web's strands and nodes.

Once the Universe cooled enough for electrons to recombine with ions and so form neutral atoms, the photons could no longer mess with the ordinary matter. Atoms of hydrogen, helium, and other light isotopes were finally free to flow towards the gravitating centers first established by the dark matter. Theorists had already predicted in the late twentieth century that the ambient density and temperature of the Universe after recombination would be most favorable to forming relatively small clumps with sizes of several hundred light years and masses of about a million Suns. Any smaller and the self-gravity would be insufficient to counter the pervasive gas pressure; much larger, the odds decrease for gravitationally isolating such a big cloud out of the expanding medium.

By working with mixes of cold dark matter, ordinary matter, and dark energy (what are known as Lambda Cold Dark Matter [ΛCDM] scenarios), the numerical simulations have since produced impressive details of the emerging Cosmic Web and of the condensing galaxies and galaxy clusters therein. No single code can simultaneously cover the incredible dynamic range of spatial scales and associated structuring, so most specialize. Together, they have shown that the first galaxies should have looked like gherkin pickles rather than the disks and spheroids we see today. Perhaps that is what we are seeing in the Hubble Deep Fields and other deep surveys. The simulations have also found that spiral galaxies, once formed, are more resilient than expected. Perhaps the anticipated mash-up between the Milky Way and Andromeda Galaxies some 4 billion years from now will not be as disastrous as once thought.

Most important, the numerical simulations have discovered a strong sensitivity to energetic feedback from newborn stars in the galaxies. The most massive stars, in particular, produce intense UV radiation, strong winds, and ultimately supernova explosions that stir up the gravitating clumps of dark and ordinary matter. This strong stellar feedback may serve to retard the further formation of stars, while reducing the total number of clumps. While the standard Lambda Cold Dark Matter scenario for creating galaxies predicts way too many dwarf galaxies swarming around a giant galaxy like the Milky Way, the addition of stellar feedback seems to get rid of this issue.

Reionization of the cosmos

The numerical recipes for forming galaxies ultimately lead to the forging of stars. Before the stellar manufacturing of elements heavier than helium, however, starbirth as we know it was a tall order. By themselves, hydrogen and helium atoms at temperatures of several hundred kelvin are poor radiators. That means they cannot give up the gravitational energy that they inherited from the collapse of their birth clouds. And that means no further cooling and collapsing. Given this state of affairs, astrophysicists have predicted that the first stars had to be incredibly massive and luminous monsters. Only a very massive birth cloud would have the gravitational wherewithal to counteract the high temperatures and pressures that would remain in a condensing cloud of hydrogen and helium.

With predicted masses of several hundred Suns and UV luminosities of many million Suns, these first-generation stars would have completely ionized the ambient gas in their host galaxies. Moreover, they would have likely ionized any tenuous gas between the galaxies. Astrophysicists have dubbed this critical time the reionization epoch, and astronomers have looked for evidence of it. To date, they can see that the spectra of quasars at lookback times exceeding this epoch are degraded by absorptions from the intervening neutral atoms of hydrogen and helium, while quasars at the smaller lookback times that follow this epoch do not show such degraded spectra. That is because in the latter case the light could stream through the reionized hydrogen and helium without any further absorptions. In 2015, astronomers found what they think are actual first-generation stars in the young galaxy COSMOS Redshift7. This unusually bright galaxy was observed at a lookback time of about 13 billion years, when it was radiating mostly in the ultraviolet. Moreover, it showed no evidence of emissions or absorptions from elements heavier than helium, consistent with an unenriched population of first-generation stars.

Today, we inherit our ionized intergalactic medium from the reionization epoch. We also inherit a fair quantity of heavy elements from the first generation of stars. After their brief but brilliant lives, these stellar pioneers would have exploded as supernovae – thus injecting salvos of newly forged elements into their host galaxies and beyond. Once enriched with these heavier elements, the clouds in these galaxies could radiate away their gravitational energies, cool off, and so collapse into the more "normal" stars we recognize today.

Merging and melding

Once small galactic clumps had broken away from the unceasing expansion, they began to gravitationally aggregate into larger structures. It is worth remembering that our observable Universe 13 billion years ago was smaller by a factor of 15 or so and thus denser by a factor of roughly 15 cubed, or 3,400, than it is at present. The rate of collisions between the sub-galactic clumps would scale with the density, and so in the crowded early Universe, close interactions and mergers were most likely the order of the day.

So, what got made during this quickening time? Perhaps larger spheroids were first to form, followed by disks in those merging systems which had significant angular momentum. Astronomers don't really know. They do know, however, that giant galaxies made their debut in a startlingly brief amount of time. By about

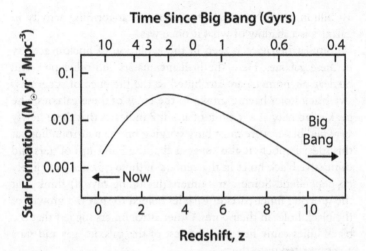

Figure 10.2 History of cosmic star formation, where the star formation rate per unit co-expanding volume is plotted as a function of redshift and corresponding lookback time. (Adapted from P. Madau and M. Dickinson, "Cosmic Star Formation History," *Annual Review of Astronomy and Astrophysics* [2013], in the NASA/IPAC Extragalactic Database.)

1.5 billion years after the big bang, galactic monsters had taken the cosmic stage. These galaxies were up to ten times more massive than our Milky Way Galaxy and ablaze with star formation. How they built themselves up so quickly remains a big challenge for astrophysicists.

Comprehensive studies of galaxies over a range of redshifts and inferred lookback times have revealed a fairly clear evolution in the average star-forming rate (see figure 10.2). Most galaxies, including the Milky Way, began forming stars more than 12 billion years ago – less than 1.8 billion years after the big bang. The star-formation rate then ramped up to a peak that is evident at a redshift of about 2, corresponding to a lookback time of about

10 billion years. Nowadays, the average star-forming activity in galaxies is a shadow of what it once was.

Similar behavior is seen in the evolution of nuclear activity in these galaxies. Here, the brilliant quasars and other powerful nuclear phenomena are attributed to the presence of supermassive black holes having grown in the cores of these galaxies. The peak of activity at a redshift of about 2 indicates that these newly spawned beasts were most busy gorging on their surroundings at this special epoch. It also suggests that the build-ups of stars and of massive black holes in the centers of these galaxies were likely co-dependent. Some astronomers (including myself) think that the growing inner parts of galaxies helped to spur the growth of the black holes in their cores. Other astronomers suggest that the black holes came first, and the rest of the galaxies' gas and stars agglomerated onto them.

Looking ahead

We now bear witness to a Universe adorned with giant elliptical and spiral galaxies, along with smaller irregular galaxies and even smaller dwarf galaxies of both irregular and elliptical/spheroidal form. Their particular properties and evolutionary histories were discussed in chapter 8, but I would like to add one more thought. Given the incessant expansion of our Universe, the spaces between galaxy groups, clusters, and superclusters will continue to increase. Within these clustered structures, however, the individual galaxies will interact with one another on time-scales of several billions to tens of billions of years. There will be winners and losers, with the total number of galaxies likely decreasing in sync with all the merging and melding.

Meanwhile, the amount of gas available to form new stars will subside with every low- and intermediate-mass star that gets made. These stars don't explode and so they sequester much of the matter that went into making them. Slowly, the galaxies will host ever fewer star-forming episodes. Unless the universal expansion reverses itself, or new sources of infalling matter emerge from the Cosmic Web or from the vacuum of space itself, the cycles of starbirth and stardeath in the remaining galaxies will play themselves out. Eventually, the galaxian theatre will go dark. Does that mean that our very existence as humans on the Sun-warmed Earth represents a very special time in the unfolding and evolving cosmos? Perhaps so. These ruminations on our distant future make even more precious the physical circumstances that have led to the formation of stars like the Sun and to planets like Earth. In the next chapter, we will delve into those natal circumstances and the subsequent processes of stellar birth and early evolution.

11
The birth of stars and planets

We had the sky, up there, all speckled with stars, and we used to lay on our backs and look up at them, and discuss about whether they was made, or only just happened — Jim he allowed they was made, but I allowed they happened; I judged it would have took too long to make so many. Jim said the moon could a laid them; well, that looked kind of reasonable, so I didn't say nothing against it, because I've seen a frog lay most as many, so of course it could be done.

Mark Twain, *Huckleberry Finn*

Be humble for you are made of earth.
Be noble for you are made of stars.

Serbian proverb

While galaxies pretty much have the same birthdates, the same cannot be said of stars. Sometime around the first billion or so years after the hot big bang, sub-galactic clumps were merging and melding to form the menagerie of galaxies that we recognize today; stars, by contrast, were forming way back then and have continued to form up to the present epoch. So, what does it take to form one of these thermonuclear powerhouses and its associated system of planets, asteroids, comets, and whatever? We can approach this question by interrogating our own Solar System, by probing the dark, dusty nebulae where protostars and protoplanetary systems are currently incubating, and by building physical models of the formative process.

Insights from the Solar System

Our Solar System is replete with salient clues regarding its formation. First, let's begin with the Solar System's birthdate. Radiometric dating of uranium, thorium, and other radioactive isotopes in meteorites has shown a clear maximum age of 4.6 billion years. The oldest lunar rocks collected and transported to Earth by the Apollo astronauts are of a similar vintage. Theoretical models of the Sun and its powering also corroborate this birthdate. By contrast, the Earth's rocky surface appears to be significantly younger. The oldest known minerals on Earth are the zircons found in the Jack Hills of Australia. These first crystallizations occurred 4.4 billion years ago. The oldest whole rocks on Earth, including those of the Canadian Shield, have ages of more like 4 billion years. So, we can conclude from these age differences that it took several hundred million years for the newborn Earth to cool enough for its outer crust to solidify.

The seminal age of 4.6 billion years also tells us that the Sun and Solar System were not the first stellar systems to form in the Milky Way Galaxy. Given the Milky Way's estimated age of 12 billion years, we can see that the Galaxy was busily forming stars for more than 7 billion years before the Solar System came along. This is an important finding, as it tells us that our planetary system was the beneficiary of thousands of generations of massive stars that had previously formed, fervently lived, and violently expired in the form of supernova explosions. Their effluvia of heavy elements then blended into the interstellar medium, from which our birth cloud gravitationally congealed. That is why you and I live on the watery surface of a rocky planet rather than somewhere within the atmosphere of a planet made entirely of hydrogen and helium. As Carl Sagan so famously said, "We are made of star-stuff."

The basic configuration and motions of our Solar System also tell us a lot regarding the genesis of planetary systems. Consider the fact that all the known planets in the Solar System – including the eight major planets, the five or more dwarf planets in the outer Solar System, and the myriad asteroids occupying the inner Solar System – all rotate around the Sun in the same direction and approximately in the same plane. With the exception of Venus and Uranus, most of the planets also spin about their axes in the same direction as their orbits. These commonalities have allowed astrophysicists to imagine the Solar System – and any other planetary system – as arising from the monolithic collapse of a slowly spinning birth cloud. This basic picture of stellar and planetary origins will be discussed in greater detail later on in this chapter.

Insights from within dark nebulae

Before the twentieth century, astronomers still puzzled over the patches of darkness with which the gauzy Milky Way is riven. Even the great astro-photographer E. E. Barnard wondered whether the patches represented obscuring clouds or actual holes in the stellar firmament. He eventually concluded in 1919 that his deep photographs had captured clouds whose content attenuated the background starlight. However, it wasn't until the 1930s that astronomers had convincing evidence for this intervening medium of gas and dust – what we now call the interstellar medium. By observing stars with a spectroscope, they could see particular spectral absorption lines that could not be attributed to the stars themselves. These narrow lines of calcium and other elements had to come from some tenuous medium that was foreground to the stars. Moreover, they found that the more distant the star was, the more profound were its spectroscopic absorptions. Similar behavior was observed in the star's overall light, in terms of its brightness and color. Something out there was both attenuating and reddening the starlight.

Key clues to the nature of this interstellar medium came with deep photographic images of the dark and bright nebulae – including the Barnard objects, named in honor of E. E. Barnard. For example, images of the well-named Horsehead Nebula (Barnard 33) showed dark nebulosity in the shape of a horse's head against a background of bright nebulosity. Astronomers were able to surmise that this nebula contained regions of both glowing gases and obscuring grains of dust.

By the 1970s, radio astronomers had developed dish antennae that were smooth enough and detectors that were sensitive enough to operate at millimeter wavelengths. Equipped with this new capability, they pointed their telescopes at their favorite nebulae in anticipation of detecting emission from whatever molecules might reside there. They were not disappointed. Spectral lines of carbon monoxide, cyanogen, formaldehyde, and other simple organic molecules were evident in abundance.

The most recent mappings of these molecular clouds show that they range in size from a few light years to hundreds of light years and in mass from a few thousand Suns to millions of Suns. Moreover, they appear to be arranged into sinuous filaments, perhaps under the influence of weak magnetic fields that thread through the interstellar medium. These filaments may play important roles in mediating the formation of cloud cores inside of them and of the protostellar systems that develop inside the cores.

Careful spectroscopic observations at radio wavelengths have shown that molecular clouds are incredibly cold, no more than a few tens of degrees Celsius above absolute zero. Their cryogenic states provide another important clue to star and planet formation. Within these frigid realms, gravity can have the upper hand over the random motions of the sundry molecules. And once gravity takes over, material condensations can ensue, thus producing molecular cores with densities more than a thousand times greater than in the ambient clouds. Indeed, astronomers have wondered why so many molecular cloud cores remain propped up against their gravitational inclination to collapse in one big starbirthing orgy. That's where the fine-tuning comes in. There are other motions to consider, including the core's rotation and internal turbulence. And the magnetic fields may also play a role, as they will intensify if they are bound to the condensing clouds. Lastly, any newborn stars will inject radiative and mechanical energy into their remnant clouds and so will provide prophylactic influences against further condensations. All of these mediating factors have been observed in molecular clouds. Their relative importance, however, remains unsettled.

Observations of the dense molecular cores have benefited from recent imaging and spectroscopic campaigns at mid-infrared wavelengths. At these wavelengths, both the complex organic molecules and microscopic dust grains profusely glow. The Spitzer Space Telescope was especially adept at mapping and characterizing the star-forming cores in the Taurus, Orion, Cepheus, and other molecular clouds that inhabit the Milky Way. Spitzer also obtained detailed images of several giant molecular clouds that host the most massive, hot, and powerful newborn stars. These nebular behemoths sport vast cavities with strange fingerlike protuberances that all point back to the exciting stars. Such excavations result from the intense scouring and shocking that the hot stars' UV light and winds have inflicted upon their nebular surroundings. The iconic "pillars of creation" that characterize the Eagle Nebula (M16) and the Soul Nebula (W5) testify to the transformative impact of massive hot stars on their natal environment. Evidence has accrued in favor of the Sun and Solar System having formed inside one of these "starbursting" clouds. That means we were likely beneficiaries of heavy elements from nearby supernovae as well as from the overall interstellar mix.

Zooming into the molecular cores, astronomers have perceived several manifestations of stellar genesis. They have detected the protostars themselves, along with any protoplanetary disks that they might have surrounding them. The Hubble Space

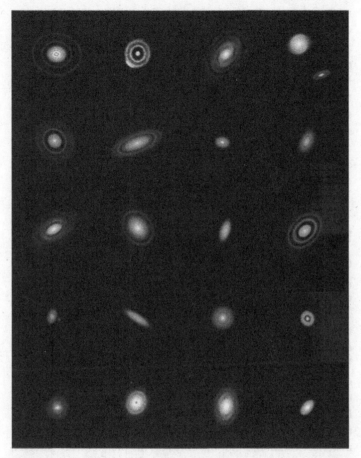

Figure 11.1 Gallery of twenty protoplanetary disks, as imaged by the Atacama Large Millimeter Array at millimeter wavelengths. Here, the light is from microscopic grains of dust at cryogenic temperatures of about 20–40 kelvin. (Courtesy of ALMA [ESO/NAOJ/NRAO], S. Andrews et al.; [NRAO/AUI/NSF], S. Dagnello.)

Telescope was the first to clearly image such disks – in silhouette against the roseate glow of the Orion Nebula. But more recently the Atacama Large Millimeter Array of radio telescopes in Chile has imaged an amazing menagerie of protoplanetary disks, some smooth, some with spiral density waves coursing through them, and some with dark rings thought to denote where forming planets have swept up the material at those radii (see figure 11.1). Lastly, multiple ground-based and spaceborne observatories have traced the immediate repercussions of stars in formation. Jets of glowing gas can be observed streaming from opposite ends of some protostars. These jets are officially dubbed Herbig–Haro objects in honor of George Herbig and Guillermo Haro, the astronomers, from America and Mexico respectively, who were the first to advance their cause back in the 1940s.

Insights from physical models

Given the observed presence of molecular cloud cores, proto-planetary disks, and mature planetary systems – including our own Solar System – astrophysicists have made major strides in delineating the overall process of star formation. Much is owed to Pierre-Simon Laplace, who first posited the nebular hypothesis for the Solar System's formation back in the early 1800s. By considering a cloud (or nebula) under the influence of its own gravity, and allowing for some overall rotation, he saw that the cloud would collapse preferentially along its rotational axis. Material along the equator would collapse less, because much of the inward-directed gravitational force is being used to bind the material's rotational motion instead. This preferentially directed gravitational collapse would itself produce a central concentration along with a flat disk of remnant material – what ultimately would become a host star along with a coterie of planets that all have the same sense of orbital motion around it (see figure 11.2).

Since Laplace's day, astronomers have wrestled with many issues associated with this hypothesis. One major conundrum is how the congealing, rotating, and flattening molecular core deals

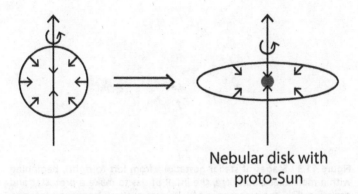

Nebular disk with proto-Sun

Figure 11.2   Cartoon of a rotating cloud that collapses via its own self-gravity into a central mass and enveloping disk, as first articulated by Pierre-Simon Laplace in the early nineteenth century. The cloud's rotation reduces the acceleration of infall at the cloud's equator, thus producing a flattened disk. The central mass will become a self-luminous star, while the disk will break up into several planets.

with its initial angular momentum (mass in rotational motion). Since the central protostar gets most of the infalling material, it should have most of the system's angular momentum. And since it collapses by many orders of magnitude, it should spin up to incredible speeds in order to conserve its initial angular momentum. As remarked earlier, similar behavior can be seen when a spinning skater brings in his arms, or when a diver curls up during her dive. This rotational dynamic is not observed, however, in collapsing cores. Consider the Sun. It contains more than 99 percent of the Solar System's mass and yet it spins around at a placid rate of one rotation every twenty-seven days. Consequently, most of the Solar System's remaining angular momentum is contained in the orbits of the giant planets – especially Jupiter. Where did the rest of the primeval Solar System's angular momentum go?

One possible solution to this angular momentum problem is to get rid of significant quantities of the original disk's spinning

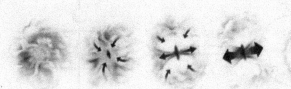

Figure 11.3 Stages of stellar gestation (from left to right), beginning with a molecular cloud core, the infall of gas to make a protostar and accretion disk, the emergence of a bipolar outflow from the protostar, the resultant clearing of material in the protoplanetary disk, and the residual star with orbiting planets that characterize the Solar System and other "mature" planetary systems. (Courtesy of Charles Lada [Harvard-Smithsonian Center for Astrophysics] and Rob Wood [Illustrator].)

and infalling mass. That means a forming solar system would have to create a major outflow away from its disk. Such a solution also would help to explain the bipolar outflows that have been observed streaming from many protostars.

Given the contending dynamics of gravity, rotation, magnetic fields, radiation, and other influences, the formation of stellar and planetary systems from nebular matter can be a bit complicated. A few of the key gestational stages thought to play out during the transformational process are depicted in figure 11.3.

The entire transformation from a recognizable protostar into a Sun-like star is thought to take only 30 million years. The realization of a 40-solar mass star (such as that powering the Orion Nebula) would take only 100,000 years, equivalent to the approximate age of humans, while the formation of a 0.1-solar mass star (such as the M-type dwarf Proxima Centauri, our next closest star) would take up to a billion years. The subsequent lives and deaths of stars also critically depend on their initial masses, as we shall see next.

12

Cycles of life and death among the stars

Not just beautiful, though – the stars are like the trees in the forest, alive and breathing.
  And they're watching me.
                                        Haruki Murakami, *Kafka on the Shore*

Nothing lasts forever, not even stars. From their cryogenic births to their vociferous youths, stable adulthoods, bloated middle-ages, and ultimate expirations, stars undergo many transmogrifications. Astronomers have been able to piece together these cycles of life and death among the stars by considering each observed star as a "snapshot" in its respective life story. A star's mass turns out to determine much of its narrative in terms of its overall lifetime and the evolutionary changes it undergoes.

Low-mass stars

Stars of lowest mass (0.08 to 0.8 solar masses with M spectral types) change the least. That is because they are fully convective (see figure 12.1). Over time, every atom in the star will cycle through the thermonuclear core, thus feeding the star's power-house until all the hydrogen fuel fuses into helium. What the end product might resemble remains unknown, because these

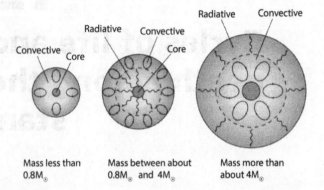

Figure 12.1 Internal structures of stars with differing mass. Low-mass stars (left) are fully convective. Intermediate-mass stars (middle) have their convective zones above their radiative zones. High-mass stars (right) have their radiative zones above their convective zones. (Adapted from *Astronomy* by C. J. Peterson.)

stars live for so long. Indeed, every low–mass star to have formed in the Milky Way over its 12-billion-year existence is still very much alive today.

Sun-like stars

Stars like the Sun (0.8 to 1.4 solar masses with K, G, and F spectral types) spend 90 percent of their lives on the relatively stable main sequence (see figure 12.2). During this time, hydrogen fusion dominates the thermonuclear core. The resulting energy that is released is then transferred outward through the rest of the star. Just beyond the core, the high–energy photons heat the gas particles which then re-radiate according to their temperatures. As the radiation wends its way outward, increasing numbers of gas particles are heated, but to ever lower temperatures compared to that in the core. In the radiative zone, what were once a few gamma-ray photons get converted to a torrent of relatively lower-energy ultraviolet and visible-light photons. Two-thirds of the way out from the core to the surface, giant convection currents take over, lofting the heated particles to the surface where they radiate (mostly at visible wavelengths), cool, condense, and then descend back down to repeat the convective cycle.

Consider the Sun itself (see chapter 5). Forged from nebular matter 4.6 billion years ago, it is midway through its life as a normal main-sequence star. During this time, hydrogen nuclei in the Sun's core have been busily fusing into helium nuclei along with neutrinos and gamma rays – the latter being the source of the Sun's luminous power. All that thermonuclear fusion has created a core increasingly made of helium nuclei, each consisting of two protons and two neutrons, rather than the original cohort of hydrogen nuclei made of single protons. And that means the total number of autonomous particles in the core has been declining steadily.

According to the ideal gas law, the internal pressure depends on both the number of particles and their collective temperature. So, if the particle number decreases, the temperature must increase in order to maintain the pressure that is necessary to fend off gravitational collapse. The result is a star whose core temperature and corresponding luminosity have been steadily increasing. The early Earth likely bore witness to a significantly dimmer Sun than is present today. In another few billion years, the Sun will be significantly brighter than it is now, and life on Earth will become untenable – even before the Sun has transitioned into its red giant phase.

Now consider the segregation between the inner radiative layer and outer convective layer in Sun-like stars. On account of this, the thermonuclear core does not have access to all of the star's hydrogen. Eventually, the core will run out of fuel, go dormant, and gravitationally contract, thereby inducing hydrogen

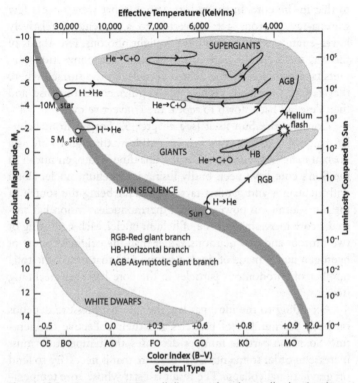

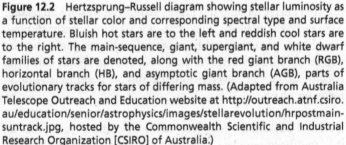

Figure 12.2 Hertzsprung–Russell diagram showing stellar luminosity as a function of stellar color and corresponding spectral type and surface temperature. Bluish hot stars are to the left and reddish cool stars are to the right. The main-sequence, giant, supergiant, and white dwarf families of stars are denoted, along with the red giant branch (RGB), horizontal branch (HB), and asymptotic giant branch (AGB), parts of evolutionary tracks for stars of differing mass. (Adapted from Australia Telescope Outreach and Education website at http://outreach.atnf.csiro. au/education/senior/astrophysics/images/stellarevolution/hrpostmain-suntrack.jpg, hosted by the Commonwealth Scientific and Industrial Research Organization [CSIRO] of Australia.)

fusion in a shell just beyond the contracted core. This hydrogen shell burning characterizes the red giant phase, in which the star's outer layers expand until the star is 100 times its former size. In less than 1.2 billion years after the end of its main-sequence stage, the contracting core will reach a density and temperature high enough for its helium to fuse into carbon and oxygen. The release of energy per fusion reaction is then considerably less, and so the rate of reactions must ramp up to keep the star from collapsing. The star is now in its horizontal branch stage (see figure 12.2), which will last only about 100 million years.

Once the helium fuel in the core is exhausted, the core will once again contract until helium begins fusing in a shell surrounding the core. That shell, in turn, will be surrounded by a hydrogen-fusing shell. The star's luminosity will climb as the star enters its asymptotic giant branch stage (see figure 12.2). At this point, the star could be as big as the orbit of Mars, and the outermost atmosphere will be sufficiently cool for certain gases to crystallize into microscopic grains of dust. What were once free atoms of carbon, silicon, and oxygen will settle down into silicate and graphite grains the size of smoke particles. Instabilities involving the two fusing shells will induce pulsations that will drive strong winds. In these winds, the dust grains are propelled outward, infusing the interstellar medium with enough dust to aggregate into planetesimals and eventually planets. That means our home planet and other rocky planets owe their respective origins to the mighty winds of once-giant stars.

For a Sun-like star, the asymptotic giant branch stage lasts a mere 20 million years. During this brief period, the star's powerful winds will remove more and more mass – thus exposing the remnant core of carbon and oxygen. That core will continue to contract until it becomes a white dwarf that is propped up against its crushing self-gravity by repulsive forces

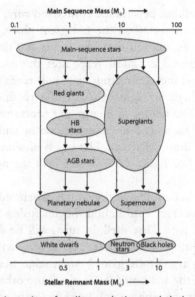

Main Sequence Mass ($M_\odot$) ⟶

Stellar Remnant Mass ($M_\odot$) ⟶

Figure 12.3 Trajectories of stellar evolution and their dependence on initial mass. Intermediate-mass stars produce white dwarfs, while high-mass stars produce either neutron stars or black holes. (Adapted, with changes, from *Discovering the Universe* by W. J. Kaufmann and N. F. Comins, 4th edition, W. H. Freeman [1996].)

between its electrons (more on this in chapter 13). Because the surface of the carbon–oxygen white dwarf is extraordinarily hot – between 30,000 and 100,000 kelvin – it produces lots of UV radiation which ionizes and fluoresces the gases vented by the winds. We now have a planetary nebula, whose exquisite form and array of colors will last a fleeting 10,000 years before dissipating into space. Without anything to change the static equilibrium of the dense stellar remnant, the white dwarf will slowly cool via conduction and radiation over billions of years (see figure 12.3).

Other intermediate-mass stars

Stars significantly heavier than the Sun (1.4–8.0 solar masses) pass through the same evolutionary phases as Sun-like stars, but with a bit of a twist. The main-sequence phase is still characterized by fusion of hydrogen into helium in the star's core. However, the fusion makes use of available nuclei of carbon, nitrogen, and oxygen to make the reactions proceed at a faster rate. These kinds of catalytic reactions require higher temperatures in the thermo-nuclear cores, which is why only the higher-mass stars can host them. The so-called CNO cycle helps higher-mass stars to pro-duce much greater luminosities during their relatively stable main-sequence phases.

High-mass stars

The dividing line between intermediate-mass and high-mass stars is thought to be roughly 8.0 solar masses. Above this thresh-old, the succession of fusion reactions in the star's core over its lifetime can go beyond the sequence of hydrogen to helium to carbon to oxygen that intermediate-mass stars can muster. The high-mass stars (8.0–120 solar masses) can host sufficiently high core temperatures for oxygen to fuse into silicon, and then for silicon to fuse into iron. Along the way, the star inflates into a supergiant whose size can exceed the orbit of Saturn. After hav-ing fused helium into carbon, oxygen, and neon, the star will fuse these elements into silicon and then iron in about a day. That's when the gig is up.

The iron nucleus has greater (negative) binding energy than all the other elements. Fusion of iron nuclei into heavier nuclei would entail *adding* (positive) energy to initiate the reaction. Such endothermic reactions work OK when there is a handy source of energy – such as the sunlight that drives the photosynthesis reactions in plants. However, in the cores of stars there is no other reserve of energy to ignite the iron. The now-dormant core gravitationally collapses in less than a second, with the resulting release of gravitational energy driving a stupendous explosion in the rest of the star.

When a high-mass star goes supernova, it can outshine the entire galaxy that it inhabits. The supernova will then dim to obscurity over a period of weeks to years. The material that it blasted into space forms a supernova remnant containing all the heavy elements that were cooked up before and during the explosion. Indeed, we have such stellar pyrotechnics to thank for much of the periodic table of the elements.

What becomes of the collapsed remnant cores of high-mass stars is discussed in the next chapter.

13

Conundrums of matter and energy

Hamlet: *There are more things in heaven and earth, Horatio, than are dreamt of in your philosophy.*

William Shakespeare, *Hamlet*

Amid the myriad marvels of planetary, stellar, and galactic pedigree lurk even more mysterious and bizarre phenomena. From impossibly dense nuggets of matter to faintly rippling waves of spacetime and ghostly apparitions of dark matter and dark energy, the cosmos continues to allure and elude the greatest of minds. These conundrums of matter and energy include white dwarfs, neutron stars, and pulsars – and, of course, both stellar and galactic black holes. We know that neutron stars exist, as we have observed them in the centers of supernova remnants – often in the guise of pulsars. We also are fairly confident that stellar black holes exist, because we have found normal stars in close binary pairings with objects of qualifying mass and invisibility. Indeed, gravitational waves were first detected in 2015 from the collision of two stellar black holes. These long-sought imprints of rippling spacetime were followed in 2017 by a detection of gravitational waves from two colliding neutron stars. The recent successes of gravitational-wave astronomers provide a stark contrast to the continuing frustrations of physicists endeavoring to understand the nature of dark matter and dark energy. There is compelling evidence for both forms of matter–energy pervading the cosmos. We just don't yet know what comprises them.

White dwarfs

As noted in the previous chapter, intermediate-mass stars (0.8–8.0 solar masses) progress through the sequence of fusing hydrogen into helium, helium into carbon, and carbon into oxygen. Further fusion reactions require greater central temperatures than can be mustered at these stellar masses. That means the fusion reactions eventually peter out, and the once-thermonuclear core contracts under its own weight. The result is a strangely beautiful nugget of carbon and oxygen with the mass of the Sun but the size of Earth. The nugget is "metallic," in that the crystallized atomic nuclei share the same bath of conducting electrons, and at a million grams per cubic centimeter it is denser than anything that we can make in the laboratory. Indeed, a teaspoon of white dwarf would have the same mass as a mid-size automobile.

The white dwarf's strangeness doesn't end with its metallic properties and astoundingly high density, however. Bizarre quantum effects begin to take hold when the atoms are so crowded together. According to the Pauli exclusion principle, no two particles can share the same exact quantum state. In a white dwarf, that means no two electrons can share the same energy. The consequences are electron energies and corresponding pressures that depend only on the density of the stellar remnant – temperature no longer plays a role in pressurizing it. This odd state of affairs, known as electron degeneracy, is what keeps the white dwarf from collapsing further.

Imagine a normal star. If you added more mass to it, the central pressure would rise, the temperature and luminosity would increase accordingly, and the star would expand in response. You can see this behavior with the stars that occupy the main sequence of the Hertzsprung–Russell diagram. Main-sequence stars of greater mass and corresponding luminosity (such as Spica, Vega, and Sirius) are significantly larger than their lower-mass counterparts (such as the Sun and Barnard's Star) (see figure 7.6). In a white dwarf, however, the same addition of matter causes the remnant to actually shrink! That is because the added matter neither increases the internal pressure nor the corresponding temperature. Instead, the remnant contracts under its own weight. There is a limit to this behavior, of course, when enough mass accumulates to produce a remnant with essentially zero radius. The great Indian astrophysicist Subrahmanyan Chandrasekhar calculated in 1930 that the limiting mass would be exactly 1.4 times the mass of the Sun. Observations of white dwarfs have yielded estimated masses ranging from 0.17 to 1.33 Suns, which would seem to confirm this limit. Most white dwarfs have masses of 0.5–0.7 Suns. If the stellar core has a mass that exceeds the Chandrasekhar limit, it is fated to form something else entirely different, as will be explored shortly.

White dwarfs may be small, but they can episodically outshine their stellar relatives by huge amounts. Because they have stellar masses concentrated into planetary-sized packages, their surface gravities are extraordinarily intense – more than a billion times the surface gravity of Earth. Anything that falls onto the surface of a white dwarf will make a big bang. That's exactly what happens when a white dwarf resides in a close, binary star system. As the partnering star takes its turn evolving into a red giant, its outer atmosphere falls under the gravitational spell of the white dwarf. Material from the hapless giant star will then stream around the white dwarf and onto its surface. At some critical moment, the added material reaches a threshold of mass and density whereby the white dwarf's surface undergoes a thermonuclear runaway of chain reactions. This bush fire of stellar proportions has been identified with the nova phenomenon, where the partnering star appears to suddenly brighten by factors of 10,000 to 16 million

(10–18 magnitudes). We now know it's the otherwise invisible white dwarf which has brightened so drastically. In the Milky Way, approximately ten novae are observed each year, with about one of these achieving naked-eye status.

The last and final act that a white dwarf can play out is to completely explode into the cosmos. This is thought to occur in close binary systems when a white dwarf accretes enough mass from its sibling star that it just exceeds the 1.4-solar mass Chandrasekhar limit. The sudden implosion of the white dwarf sets off shock waves that tear the remnant asunder. Astronomers observe the repercussions as supernovae of Type Ia, where the spectra of the outbursts show very little line emission from hydrogen – because the exploding white dwarf is mostly bereft of any hydrogen shell. By contrast, the collapsing cores of hydrogen-rich massive stars produce supernovae of Type II, which show abundant hydrogen-line emission. Because Type Ia supernovae all arise from remnants of pretty much equal mass, they all produce the same total luminosity, and that means they can be used as standard candles for determining the distances to their hosting galaxies. In this way, astronomers have established reliable distances to galaxies extending out as far as a billion light years from Earth (see figure 8.5).

Neutron stars

A massive star with a dormant core whose mass exceeds 1.4 Suns will gravitationally implode into either a neutron star or a black hole. As previously noted, the implosion will drive an explosion of the surrounding star that is observed as a supernova of Type II. Astronomers believe that the dividing line between the fates of neutron star and black hole lies somewhere around 3 solar masses. For a neutron star, the crushing stops when the remnant achieves nuclear density, the point at which the degenerate electrons have combined with protons in the carbon and oxygen nuclei to form neutrons. We now have an object equivalent to one giant nucleus composed solely of degenerate neutrons. The remnant core achieves this state after having collapsed to the size of a city – about 25 kilometers across. The resulting densities of 10^{14} grams per cubic centimeter exceed those of a white dwarf by a factor of 100 million. The corresponding teaspoon of neutron star would have a mass equivalent to that of Mount Everest – if that nuclear teaspoon was dropped onto the surface of Earth, it would plunge through all that rock like a bullet through air, and then proceed to oscillate from the point of impact to the opposite side of Earth and back like a yo-yo.

Astronomers first hypothesized the existence of neutron stars as the imploded remnants of massive stars during the 1930s, shortly after the discovery of neutrons. It wasn't until the 1960s, however, that they obtained persuasive evidence for these incredible objects. Graduate student Jocelyn Bell (later Bell Burnell) and her thesis advisor, Anthony Hewish, first noted in 1967 something odd with the radio signals being recorded at the Mullard Radio Astronomy Observatory in England. Something out there was blinking on and off with an extraordinarily regular cadence. Follow-up observations by multiple radio observatories subsequently revealed tens of blinking radio sources across the sky with pulses on timescales of seconds to milliseconds. First jokingly dubbed LGM for "Little Green Men," these pulsating sources were later given the more respectable appellation of pulsars, which has stuck ever since.

Astronomers already knew that some stars physically pulsate, but those pulsations occurred on timescales of hours to days to weeks. No normal star would be able to pulsate on second-to-millisecond timescales; it would tear itself apart in a few heartbeats. Even a white dwarf would not be able to hold itself together. A rotating white dwarf could conceivably beam radiation our way with every turn of the remnant. However, a white dwarf's gravity

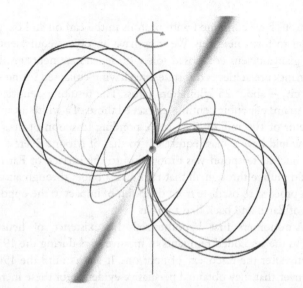

Figure 13.1 Schematic drawing of a rapidly spinning neutron star beaming electromagnetic radiation along its magnetic axis. The pulsar phenomenon is thought to result from this highly condensed, magnetized, and energized configuration. (Courtesy of Roy Smits, Wikimedia Commons.)

would be insufficient to bind it against the rapid spinning that is necessary to explain the pulsar phenomenon. What's left is the neutron star model posited back in the 1930s. With a mass significantly greater than the Sun stuffed into a volume equivalent to that of Lake Baikal in Siberia, a neutron star would have sufficiently strong self-gravity to counter its breakneck spinning.

The trick to explaining the pulsar phenomenon is to let the poles of the neutron star's intense magnetic field be directed slightly away from the star's spin axis. The same skewed arrangement between the magnetic and spin axes can be found to a lesser degree here on Earth. In this way, the magnetic poles sweep circles around the spin axes (see figure 13.1). The open magnetic field lines at the poles allow for the release of electromagnetic radiation which then sweeps past parts of the Galaxy like the beam of a lighthouse. If the Earth happens to be in the crosshairs of this beam, its radio telescopes receive a regular pulse of emission. The upshot is that astronomers continue to explain pulsars as neutron stars whose magnetic axes periodically swing through our line of sight to these sources. Because the model works like no other, it is widely accepted.

Most neutron stars do not display the pulsar phenomenon, as their bipolar beams of radiation do not pass through our particular line of sight. Nevertheless, some of these extremely small objects have been directly detected residing in the centers of young supernova remnants. Still fantastically hot from their implosive births, the freshly forged neutron stars copiously emit X-rays. The supernova remnant Cassiopeia A is an excellent example of a non-pulsar neutron star. Only 330 years old as observed by us, Cas A and its neutron star were most recently imaged by the Chandra X-ray satellite in 2017 (see chandra.harvard.edu/photo/2017/casa_life/).

Like white dwarfs, neutron stars can get very "loud" when in close binary pairings with more normal stars. Once again, the partnering star's outer atmosphere falls victim to the neutron star's intense gravity. This time, however, the stakes are much higher, as the neutron star has 300,000 times more surface gravity than a white dwarf. The resulting explosions can explain some of the most titanic releases of energy in the cosmos. These most extreme of the so-called cataclysmic variables radiate their alarm at both X-ray and gamma-ray wavelengths.

Black holes

The generic definition of a black hole is a region in space within which nothing can escape, not even light. In other words, to escape from inside a black hole, you would have to travel faster than the speed of light. Such speeds are thought to be physically impossible.

Any hunk of matter can turn into a black hole. The hard part is squeezing that hunk so that all of its matter fits within the black hole's event horizon – where the gravitational energy just barely binds the kinetic energy of any photon of light. For a non-rotating black hole, the event horizon corresponds to the Schwarzschild radius (R_S), which can be formulated as:

$$R_S = (2 \times G \times M) / c^2$$

where M is the mass of the hunk, G is Newton's constant of universal gravitation, and c is the speed of light. If you wanted to turn the Earth into a black hole, you would have to squeeze it until it fit inside a radius of nine millimeters – equivalent to that of a small marble. Preventing that outcome are the strong electrochemical bonds that characterize our planet's rocky interior. The Sun has a corresponding Schwarzschild radius of three kilometers, which it will never achieve as it is insufficiently massive to collapse beyond its Earth-size white dwarf state.

Stellar black holes

Dormant stellar cores with masses of 3 to 30 solar masses can collapse beyond the white dwarf and neutron star states to become black holes. That is because the degenerate electrons in a white dwarf and the degenerate neutrons in a neutron star cannot generate sufficient pressure to counter the crushing gravity of these more massive cores. Instead, the dormant cores are thought to directly implode past their corresponding Schwarzschild radii of 9–30 kilometers and so disappear inside their respective event horizons. What goes on inside the event horizon is anybody's guess. Does the imploding matter achieve the singularity, where all the mass is contained within a null volume? This nonsensical state would imply an infinite density, which most physicists decry. Perhaps the intensely warped spacetime inside the event horizon would make this dilemma moot? What physicists do tell us is that whatever is going on within the zone of unknowing, the black hole will still manifest a mass, charge (if any), and angular momentum. The mass, in particular, provides the most fundamental means for astronomers to infer the presence of stellar black holes.

Consider spectroscopic observations of an erstwhile normal star. If the star's spectral absorption lines show periodic Doppler shifts in wavelength, astronomers can infer that the star is part of a binary system that includes an invisible partner. And if the estimated dynamical mass of that partner exceeds 3 solar masses, then the object is most likely a stellar black hole.

Another way to infer the presence of a stellar-mass black hole is to detect and then monitor variable sources of X-rays and gamma rays. If the source can be identified with a visible star, then that star is likely in a closely interacting binary system with a white dwarf, neutron star, or black hole. The variability and energetics of the emission will provide further important clues for discriminating among these three possibilities. The Rossi X-ray Timing Explorer, launched in 1996, was especially adept at tracking the theatrics of these vociferous objects across the sky (see Recommended Reading and Resources).

About twenty stellar black hole candidates are known in the Milky Way, but it is likely that a whole lot more populate our Galaxy. Estimates based on the total number of stars in the Milky Way and the fraction of them sufficient in mass to spawn a black hole would suggest about 100 million stellar-mass black holes inhabiting the Galaxy's disk, bulge, and halo. A full listing of stellar, intermediate-mass, and supermassive black hole candidates can be found in the *Black Hole Encyclopedia* which is curated by McDonald Observatory (see Recommended Reading and Resources).

BLACK HOLES ARE NOT RAVENOUS SUCKING MACHINES

One of the major misconceptions surrounding black holes is that their gravitational influence extends much farther than that of other forms of matter. The popular picture of a black hole inhaling everything in sight is mostly erroneous. For example, if you could somehow replace the Sun with a solar-mass black hole, pretty much nothing would change. All the planets, asteroids, and comets would continue in their respective orbits around the black hole. The Solar System would be much darker, of course, so life on Earth and other planets or moons would have to revert to internal sources of heat and chemical energy. Meanwhile, to feel the weird effects of warped spacetime, one would have to travel well within the Sun's original radius. Then and only then would Einstein's predictions of time dilation, length contraction, tidal stretching, and severe light bending come to the fore.

Intermediate-mass black holes

Astronomers have spent decades trying to find black holes with masses between those of stellar remnants and the supermassive black holes that have been found at the cores of giant galaxies. Such objects would have masses of hundreds to many thousands of solar masses. Success has come just recently, and so far there are only six candidates listed in the *Black Hole Encyclopedia*. Three of those reside in the centers of globular clusters, while the others occupy the centers of low-mass galaxies or the disks of higher-mass galaxies. Recently, an international team of astronomers has claimed another few dozen middle-weight candidates. Elucidating the population of intermediate-mass black holes with respect to their galactic hosts is important, because it will help astrophysicists determine how the black holes got made. Did they start as stellar-mass "seeds" which then accreted other nearby stellar black holes, or did they form in one big collapse of some dense primordial cloud?

Supermassive black holes

The best evidence for black holes can be found in the centers of large galaxies. Here, supermassive black holes containing millions to billions of solar masses have been inferred from high-resolution spectroscopy of the innermost galactic regions. The spectroscopy indicates stars in exceedingly fast orbits which can only be bound if there is an enormous gravitating mass interior to them. Thanks to the high spatial acuity of the Hubble Space Telescope, stars within only a few tens of light years from the galaxies' centers have been clocked. The inferred central masses, often combined with associated nuclear activity, have built a strong case for supermassive black holes occupying the cores of most large galaxies.

Astronomers have also found an important relation between the masses of the central black holes and the masses of their surrounding galactic bulges. The black hole-to-bulge mass ratio appears to cluster around 1:1,000. This consistent correlation implies that the black holes and galactic bulges likely evolved together. The details of their co-evolution remain uncertain, however.

The most compelling evidence for supermassive black holes can be found in the center of our own Milky Way Galaxy. That is because the Galactic center is more than ninety times closer than any other giant galaxy's center and so can be studied in much finer detail. Using the giant Keck telescopes atop Mauna Kea on Hawai'i, the Galactic Center Group at UCLA, under the leadership of Andrea Ghez, has tracked the Galaxy's innermost stars for more than a decade. The star S0-2, in particular, gets to within 0.0019 light years (120 AU) from the Galactic center. That distance is equivalent to the orbital radii of some trans-Neptunian objects in the outermost reaches of our Solar System. S0-2's blistering speed of 2.5 percent the speed of light at this distance informs astronomers that a supermassive black hole of 4 million solar masses dwells within. (Ghez and Reinhard Genzel received the 2020 Nobel Prize in Physics for this achievement.)

More recently, the UCLA group has determined the three-dimensional kinematics of S0-2's orbit. This additional information has shown that S0-2 is following a trajectory consistent with Einstein's relativistic theory of gravity rather than Newton's inverse square law. Einstein's warping of spacetime in the presence of such concentrated matter continues to provide the best model for understanding the behavior of matter and light near black holes.

In 2019, even bigger news regarding supermassive black holes made headlines worldwide. The Event Horizon Telescope (EHT) team announced on April 10 that they had successfully imaged the supermassive black hole in the center of the giant elliptical galaxy Messier 87. By combining the radio signals from eight observatories situated around the world, the team was able to construct an image of sufficiently fine detail for the black hole to be resolved (see eventhorizontelescope.org). Astronomers interpret the ring of light as gravitationally lensed emission from material just beyond the black hole's event horizon. The team also trained their radio telescopes on the much nearer supermassive black hole in the Milky Way. On May 12, 2022, the EHT released its long-anticipated image of our Galaxy's central black hole. Though 1600 times smaller and less massive than the beast lurking within M87, the Milky Way's black hole looks remarkably similar – thus affirming the gravitational physics at work in both venues.

Gravitational radiation

During the gestation of this book, the field of gravitational-wave astronomy has transformed from an elusive quest into a valid science of tremendous import. Gravitational radiation was predicted by Albert Einstein in 1916 as a key outcome of his gravitational theory, whereby Newtonian forces are replaced by the warping of spacetime in the presence of matter. Any asymmetric perturbation in the arrangement of gravitating matter is then predicted to produce a series of outward-propagating ripples in the fabric of spacetime – a gravitational wave. The experimental hunt for gravitational radiation began in the late 1960s, but it wouldn't be until the late 1980s that sufficiently precise instrumentation began to be developed. The necessary precision is beyond belief, as the effects of a passing gravitational wave from a distant source will distort the fabric of spacetime at Earth on levels less than 1 percent the diameter of a proton!

On September 14, 2015, the twin Laser Interferometer Gravitational-Wave Observatory (LIGO) facilities in Hanford, Washington and Livingston, Louisiana simultaneously detected a gravitational-wave "chirp" lasting less than a second whose frequency rose dramatically. The source of this signal was best modeled as a couple of in-spiraling and merging black holes. The black holes had masses equivalent to 36 and 29 Suns, respectively, which places them at the high end of the stellar black hole category. For this amazing feat, twenty-five years in the making, the LIGO team received the 2017 Nobel Prize in Physics.

On August 17, 2017, the two LIGO facilities and the newly commissioned Advanced Virgo detector at the European Gravitational Observatory in Italy detected fainter but much longer-lasting gravitational radiation from two merging neutron stars. The third observatory enabled astronomers to triangulate the region of sky containing the source, so that follow-up observations with telescopes could be made across the electromagnetic spectrum. The violent merger event occurred in an elliptical galaxy 130 million light years away, producing a kilonova that could be monitored by gamma-ray and X-ray observatories. The titanic explosion released a huge cloud of neutrons, which then reassembled into a menagerie of heavy nuclei. Indeed, some astrophysicists now believe that the gold in our jewelry and the uranium in our reactors owe their origins to these sorts of neutron-star smash-ups.

The gravitational-wave signals from more than twenty merging events have been detected since the first discovery in 2015. Meanwhile, new instruments are coming online around the world. We are now at the dawn of gravitational-wave astronomy,

whose promise includes finding lots more merging and exploding compact stellar remnants, characterizing core-collapse supernovae, understanding heavy-element nucleosynthesis processes,
tracking mergers of supermassive black holes in galactic nuclei,
and perhaps someday detecting the hot big bang itself.

Dark matter

The case for dark matter in galaxies and the larger cosmos grows
ever greater. Historically, the first evidence-based proposal for an
invisible form of matter was made in the 1930s by the famously
irascible Swiss astronomer Fritz Zwicky. Working at the California
Institute of Technology, Zwicky invoked what we now call dark
matter in order to explain the high speeds of galaxies inside rich
galaxy clusters. These galaxies were swarming at velocities that
could not be gravitationally contained by the visible stars, gas,
and dust within them. Something else must be inhabiting the
galaxy clusters. Zwicky dubbed it the "missing mass." Really it's
not missing but rather is present in some strangely "dark" form.

In the 1970s, Vera Rubin and Kent Ford of the Carnegie
Institution found that the stars in individual spiral galaxies were
orbiting their respective galactic centers at speeds that could not
be bound by the visible matter in these galaxies. They too recommended that some unseen substance is providing the necessary gravitational glue. Their suggestion received further support
when radio astronomers clocked the orbits of gas clouds in spiral
galaxies out to radii well beyond the stellar limits. The resulting
rotation curves stayed insistently flat, contrary to the expected
radial falloff of speeds (see figure 13.2).

For me, the clincher in favor of dark matter pervading galaxies and the spaces between galaxies in clusters has come from the
HST's images of galaxies and galaxy clusters that happen to lie
along the line of sight to more distant galaxies. In these images

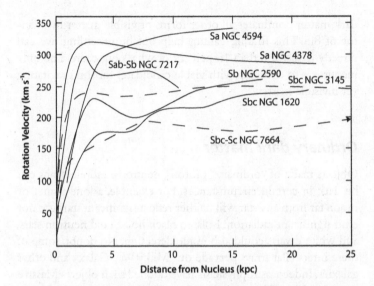

Figure 13.2 The rotation curves of nearby spiral galaxies by galaxy type show orbital velocities that remain essentially flat with increasing radius. This behavior contradicts expectations from the radial falloff of stellar light in these galaxies. Instead, the integrated masses of these galaxies are inferred to keep growing linearly with radius. (Adapted from A. Del Popolo, with reference to A. Bosma, "Nonbaryonic Dark Matter in Cosmology," *Int. J. Mod. Phys.* D23 [2014] 1430005 arXiv:1305.0456 [astro-ph.CO].)

ghost apparitions of the background objects are seen mingling among the foreground galaxies (see, for example, en.wikipedia. org/wiki/Gravitational_lens). The observed warping of the background objects can be completely explained by the effects of gravitational lensing of the background galaxies by the foreground cluster's dark matter. Analyses of these gravitational lensing effects have revealed the presence of gravitating dark matter both within the individual galaxies in the clusters and in the spaces between the galaxies. Astronomers have concluded that dark matter dominates all other forms of visible matter by a factor of 8.5. This finding cannot help but be unsettling: we can directly detect less than 15 percent of what's out there. The only recourse is to come up with viable candidates for this gravitating yet invisible matter.

Ordinary dark matter

Objects made of "ordinary" protons, neutrons, and electrons can be dark in certain circumstances. For example, a lone planet or moon far from any star will neither reflect significant starlight nor emit significant radiation. Isolated black holes, cold neutron stars, and white dwarfs could also evade detection. No doubt, some of these interstellar strays pervade our Milky Way Galaxy and other galaxies. Indeed, astrophysicists have dubbed such objects Massive Compact Halo Objects (MACHOs). The key question is whether or not there are enough of them to account for the 85 percent quota of invisible yet gravitating matter that has been inferred. So far, astronomers have yet to find planet- and moon-sized objects that are not part of the pre-planetary disks and mature solar systems that abound in our home Galaxy. Even if all the stars in all the galaxies contain planetary systems, and half of these systems have escaped into the darkness, the amount of itinerant matter would account for only a small percentage of what's needed. The same shortfall is found in estimates of the other MACHOs that have been proposed.

Another limit on ordinary dark matter comes from the nucleosynthesis that played out during the first few minutes of the hot big bang. During this time, neutrons and protons were fusing into nuclei of helium-4, along with trace amounts of deuterium, helium-3, and lithium. Single neutrons decay back into protons and electrons in about five minutes, and so all the nucleosynthetic activity had to occur before this limiting time interval. The finite duration, in turn, limited the amount of nuclear matter that could be distilled out of the chaos. In order to match the relative amounts of hydrogen, helium-4, and other isotopes that are observed in the present-day Universe, astrophysicists find that the density of all this ordinary matter amounts to only 2–5 percent of that necessary to "flatten" the fabric of spacetime. A topologically flat cosmos is one of the key findings to come out of recent mappings of the cosmic microwave background (as seen in chapter 9). To explain this geometric finding, we need lots more dark matter, together with an overwhelming quota of dark energy (see the last section of this chapter).

Hot dark matter

Non-ordinary dark matter might come in hot, warm, and cold varieties. We know that the subatomic particles called neutrinos exist in great abundance, and that they represent a form of hot dark matter due to their minuscule masses and correspondingly relativistic speeds. Such extreme speeds prevent them from being bound by individual galaxies and perhaps even galaxy clusters. They could play a gravitating role on much larger scales, but they cannot explain the dark matter that has been inferred on galactic scales. Some physicists have proposed a type of "sterile neutrino" that interacts even less with ordinary matter than the neutrinos that have been detected to date. These particles could have much higher masses and correspondingly lower velocities – low enough to be bound by individual galaxies. Multiple searches for these and other Weakly Interacting Massive Particles (WIMPs) are currently underway. The favored strategy is to place sensitive particle detectors deep underground in order to reduce the troublesome background of cosmic rays. Despite decades of effort, no experiment has yet yielded a reproducible detection.

Cold dark matter

Cosmologists are enamored with the notion of cold and hence slow-moving WIMPs, because they can explain the first hints of structure in the cosmic microwave background, the inferred presence of dark matter inside galaxies and galaxy clusters, the structuring of galaxies on larger scales, and the galaxy formation process whereby ordinary matter continuously flows along filaments towards pre-existing lumps of dark matter (as described in chapter 10). Proposed cold dark matter particles include heavier counterparts to the menagerie of particles we currently know to exist. Such "shadow" particles are predicted from supersymmetric theories of elementary matter. Selectrons, neutralinos, photinos, and gravitinos are just of a few of the supersymmetric heavyweights that have been proposed. Alas, our most powerful particle accelerators and most sensitive detectors have yet to find anything that would confirm their existence.

Modified Newtonian Dynamics

I would be remiss if I didn't mention in this discussion a competing hypothesis that would completely do away with the need for dark matter. Dubbed Modified Newtonian Dynamics (MOND), this hypothesis posits that the force of gravity from a massive object falls off with distance in a way that differs slightly from Newton's famous inverse square law (discussed in chapter 3). Alternatively, the gravitational force could remain Newtonian, but an object's response to that force – its acceleration – would differ from Newton's second law of motion, $a = F / m$. The differences would mostly become significant at very great distances. That is why the motions of the planets around the Sun appear to obey Newtonian expectations, while the motions of stars and gas clouds in the outer parts of galaxies exceed those predicted from the galaxies' observable matter and Newton's law of gravity. Astrophysicists have also invoked MOND to explain the detailed rotation curves of spiral galaxies and the strong observed relation between a galaxy's overall brightness and its rotation rate (the Tully–Fisher relation). Still, most astronomers find that dark matter can explain a whole lot more.

So, the delineation of dark matter versus its alternatives continues to elude our best efforts. We remain in a "wait and see" situation here.

Dark energy

The call for yet another invisible form of matter–energy first gained strength in the 1990s when astronomers working with data from the HST saw a slight deviation from Hubble's eponymous relation between a galaxy's distance and its redshift (as explored in chapter 8). At the greatest distances and corresponding lookback times, the observed redshifts appeared to be slightly less than expected. That meant that galaxies in the early Universe were expanding away from one another at slightly lower rates than what we see in the current epoch. Going forward in time, the universe of galaxies seems to be expanding at an ever-increasing rate. This accelerating expansion begs for an explanation. Is there some sort of "repulsive gravity" that is driving this augmented expansion? Or is something else afoot?

In order to explain the observed accelerating expansion of the galaxian Universe, cosmologists have invoked a new form of matter–energy – what they call dark energy. This same ineffable presence, if extant in large enough doses, can handily account for the exquisite topological flatness of the cosmos (see chapter 9). Moreover, scenarios of galaxy formation, evolution, and clustering seem to work best if one adopts a mix of 73 percent dark energy, 26 percent dark ordinary and exotic matter, and 1 percent luminous ordinary matter – this is the Lambda Cold Dark Matter (ΛCDM) model referred to in chapter 10.

The explanatory success of dark energy has made it very popular with cosmologists and astrophysicists. However, a big question remains – what are we dealing with here? Albert Einstein, in his cosmic field equation, required a positive energy density or "pressure" term in order to counteract the negative pressure of gravity. He included this term as an ad hoc means to keep the Universe from collapsing in on itself. Upon hearing that the Universe was expanding, Einstein described the inclusion of this term as the biggest blunder of his career. Today, we have reinstated this cosmological constant (Λ) as a vital component of the cosmos that begs understanding.

Some cosmologists think that dark energy arises from the expansion of space itself. As the cosmos inexorably expands, ever more dark energy is added to the mix of dark and luminous matter. That means the balance of dark matter and dark energy has evolved over time, with the early Universe having been dominated by dark matter (see figure 13.3), in contrast with the preponderance of dark energy in the current epoch. Some support for this evolution has been found in the record of galaxy redshifts as a function of distance and corresponding lookback time. At lookback times exceeding 10 billion years, the measured redshifts do not indicate any accelerating expansion. However, at more recent lookback times, evidence for an accelerating expansion and associated dark energy gets much stronger.

Personally, I wonder whether all the gravitationally driven condensations that have transpired over cosmic time have induced a corresponding repulsion between the condensing objects. Perhaps some sort of conservation principle involving the release of gravitational energy is at work here. Forming galaxies would be especially prone to this anti-gravitational addition to the nominal cosmic expansion (the so-called Hubble flow).

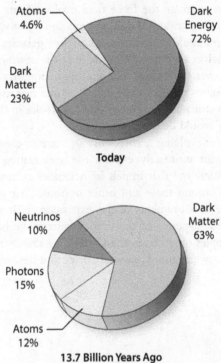

Figure 13.3 Balance of dark matter and dark energy today compared with 13.7 billion years ago. As the cosmos has expanded, increasing amounts of dark energy are thought to have added to the mix, until it is today the dominant form of matter–energy. (Courtesy of NASA/WMAP Science Team.)

This notion is highly speculative, yet it shares qualities with more established ideas by invoking some kind of agent to drive the accelerating expansion. Dubbed quintessence or "phantom energy" by cosmologists, this agent could come in the form of a force field that is spatially dependent and changes over time.

Cosmologists estimate the force field evolved from inward- to outward-trending around 10 billion years ago – just when matter began to dominate over radiation, and when galaxies had begun to get themselves together. This picture of the cosmic expansion accelerating ever faster leads to a dire prediction, however. Over unknown billions of years, our Universe could tear itself apart ... down to the atoms in our planets and the nuclei in those atoms. This 'big rip' would be our death knell.

So, will the evolving quintessence or the ever-steady cosmo-logical constant ultimately explain the accelerating expansion and the dark energy that impels it? Scientists currently cannot discriminate among these and other options. That will require more precise measurements of the expansion itself, along with more detailed mapping of the cosmic microwave background. The faint ripples observed in the CMB have already revolution-ized cosmology and may yet reveal portentous new insights.

14

Our cosmic inheritance

When we try to pick out anything by itself,
we find it hitched to everything else in the Universe.

John Muir, *My First Summer in the Sierra*

You are a child of the universe,
no less than the trees and the stars;
you have a right to be here.
And whether or not it is clear to you,
no doubt the universe is unfolding as it should.

Max Ehrmann, *Desiderata*

As corporeal creatures, each and every one of us embodies the full course of cosmic history. Every proton and neutron in every nucleus of every atom of our bodies was distilled from the seething chaos that reigned during the first millionth of a second of the hot big bang. The electrons that enfold every nucleus to make whole atoms were liberated from the fiery broth within the first second. So, in the blink of an eye, the cosmos had already spawned the subatomic constituents that are vital to our atomic and molecular selves.

The alchemy of life

The heavier atoms in our bodies have varied cosmic pedigrees. For example, the carbon in our bones and muscles came from the thermonuclear fusion of helium nuclei deep in the hearts of intermediate-mass stars during their giant, horizontal-branch phases. As the giant stars further evolved into unstable asymptotic-giant-branch stars, they released their manufactured carbon via powerful winds. The Sun is an intermediate-mass star and so will take its turn as a carbon infuser in another 4–5 billion years.

The oxygen in the water that permeates all life on Earth came from the latter stages of more massive stars, where central pressures and temperatures were sufficiently great to fuse carbon nuclei into oxygen, silicon, and iron. All these essential elements became the feedstock of subsequent stars, planets, and life when the expired massive stars ultimately exploded as supernovae and spewed their chemical creations into the Galaxy.

Rocky planets such as Earth have become the ultimate beneficiaries of this elemental largesse. Astronomers imagine that these planets grew from crystals containing silicon and carbon that had previously solidified out of the relatively cool outer atmospheres of red giants. Expelled by the stars' winds, the ice-coated crystals then wafted through the Milky Way and ultimately blended into any forming cloud of interstellar gas that was in their path. As portions of these clouds continued to gravitationally collapse into protostellar nebulae and subsequent protoplanetary disks, the crystals aggregated into microscopic grains of dust which then further agglomerated into droplet-sized chondrules, pebble-sized planetesimals – and, in a few million years of gravitational condensation, planets like Earth. The idea of Earth and all its inhabitants being "star stuff" may sound far-fetched, but it's true!

Once Earth settled down as a bona fide planet, the heat released from its formation, along with energy from radioactive isotopes of uranium and thorium, kept its interior warm and convectively dynamic, thus ensuring the plate tectonics, volcanism, and other mountain-building activities that have characterized Earth's surface for billions of years. The uranium and thorium nuclei are thought to have originated from the violent neutron-rich environments associated with in-spiraling and merging binary neutron stars. (See apod.nasa.gov/apod/ap171024.html for a periodic table of the elements with their cosmic origins identified.) Earth has its fair share of these unstable isotopes, whose continuing radioactivity helps to drive the geological cycles of crustal uplifting, weathering, deposition, sedimentation, metamorphism, and subduction. Without this tectonic foment over hundreds of millions of years, we would not have any sedimentary or metamorphic rocks to record the presence of prior life forms over this same "deep time." We would be completely ignorant of our biological past. Instead, the fossil record in these rocks has made plain that we are but the latest players in an incredibly rich heritage of biological being and becoming.

Legacies of life

The geological record of life on Earth has shown that microscopic life forms can evolve into ever more macroscopic and complex metabolizers, reproducers, and transformers of their environment. Over the past 3.5 billion years, the menagerie of living entities has grown into a vast and wondrous edifice of kingdoms, orders, families, genera, and species. What other living legacies could be inhabiting the greater cosmos? This key question continues to haunt the dreams of astronomers, geologists, chemists, and biologists alike. To date, we have yet to find any unambiguous evidence for life beyond Earth. But that has not stopped us from earnestly seeking out tell-tale signs in the meteorites, lunar samples, and planetary assays available to us. Alas, we barely know what we should be looking for.

Most searches have focused on planets and moons where water could be in liquid form and where carbon-based chemistries can take hold. Liquid water has provided the key solvent for all biochemical processes on Earth. However, other liquid solvents could be providing similar services on worlds where temperatures are much higher or lower than on Earth. A case in point is the surface of the Saturnian moon Titan, where seas of methane and ethane have been discovered in the context of a rich nitrogenous atmosphere – all at a mind-shattering temperature of $-179\ °C$. What sort of chemical transmogrifications could be (slowly) playing out in this surreal context?

Carbon-based chemical processes make up all of organic chemistry, which, in turn, underpins all life on Earth. Carbon is the leading hook-up artist, but other elements in its group on the periodic table (silicon, germanium, etc.) could also fit the bill under certain circumstances. So, we don't even know what sort of chemical signals to probe in our spectroscopic studies of exoplanets. Our current best guess is to look for evidence of methane, ozone, and oxygen molecules in the planetary atmospheres. Free diatomic oxygen (O_2), in particular, is highly reactive and so its detection would indicate biochemical processes, such as photosynthesis, that are actively replenishing the atmospheric stores of oxygen. Ozone (O_3) would be the next-best proxy for the presence of free diatomic oxygen. It has strong absorption features in the infrared part of the electromagnetic spectrum, so astronomers are keen to look for it with the next generation of giant telescopes.

Questions also arise as to the *forms* of life that could inhabit a world. Are self-contained cells prerequisite to the job of exchanging material and energy with one's environment? And how will that environment (including its gravitational, thermal, magnetic, and electric properties) constrain the life forms populating their respective worlds? Once initiated, must life always evolve towards ever greater complexity? That has been the case on Earth, but perhaps on other worlds, this trajectory might not be so inevitable.

As these questions indicate, the search for extraterrestrial life with truly open eyes is a most daunting task. Yet, cosmochemists, astrobiologists, and other interdisciplinary scientists have taken on the challenge big time. Meanwhile, the ever-increasing tally of some 4,000 exoplanets in orbit around a myriad of stars has put a face on the quest, spurring on their efforts. For us Earthlings, the search really has just begun.

Coda

It is beyond the scope of this Beginner's Guide to provide a comprehensive survey of biological evolution on Earth. However, it is within this book's wheelhouse to underscore the physical ties that bind us as humans to the greater cosmos. We all share a rich cosmic inheritance which has made us who we are today – from the creation of all matter 13.8 billion years ago, to the formation of the Milky Way Galaxy some 12 billion years ago, the birth of the Sun and Solar System 4.6 billion years ago, the emergence of microbial life on Earth 3.5 billion years ago, the evolution of complex creatures 500 million years ago, the beginning of life on land 300 million years ago, the reign of dinosaurs which began 240 million years ago and ended 66 million years ago, and the rise of hominids just a few million years ago.

We are indeed the stuff of stars, as connected to the cosmos as the Galaxy that spawned us. Yet, many key questions remain. Here are my top ten:

1. How did the diverse subatomic particles that make up all matter acquire their particular masses, charges, and spin states during the first second of the hot big bang? Was this menagerie of fundamental particles the luck of the draw or preordained?
2. What characterizes dark matter other than the incredible aloofness of its hypothesized particles? When will we ever get *direct* evidence of its existence?
3. Is dark energy a necessary and dominant component of the cosmos, or is it perhaps a phantom of astronomers' imaginations?
4. How did galaxies get themselves together so soon after the hot big bang? It seems a very tall order for the formless cosmic

plasma, with overdensities of only 1 part in 100,000, to have gravitationally condensed into giant swirling galaxies of stars and planets within just a few hundred million years.

5. What goes on inside the event horizons of black holes? Are they portals to other spacetimes or simply compacted miasmas of matter, forever cloaked from view by their own intense gravity?

6. What star-forming cloud made the Sun and Solar System? What other stellar and planetary systems in the Milky Way share our nebular provenance?

7. When, where, and how will we find the first direct evidence for microbial life beyond Earth?

8. When, where, and how will we find the first direct evidence for complex life beyond Earth?

9. When, where, and how will we find the first direct evidence for intelligent and technologically communicative life beyond Earth?

10. Pending the discovery of sentient life forms, what will be the best way to communicate with them – and should we choose to do so?

Even without considering the prospect of other universes, these questions underscore how much we still don't know about our own cosmos – and our place in it. As a technologically emergent species on the Galactic stage, we will need lots more time and concerted effort to answer these essential unknowns. Meanwhile, our own existence as part of Earth's biosphere has become perilously tenuous. Will our polluting ways end up poisoning the ecosystems upon which we rely? Will the greenhouse gases that we keep emitting warm our atmosphere and oceans beyond recovery? Will overpopulation, global starvation, and pandemics reign supreme? Or will our nuclear weaponry take us out first? These existential threats challenge us to our core. Yet I remain optimistic that we will learn to face these challenges and take the necessary steps to resolve them. The big question is a matter of timing. How bad does it have to get before we finally step up? We have just begun to take our place as viable citizens of the cosmos. We have so much to lose, but even more to gain.

After sixty years of robotic and human spacefaring, the open road to set foot on diverse worlds in the Solar System lies before us as never before. The requisite rocket technologies to get us to Mars and beyond are being built even as you read this paragraph. Once harebrained schemes for robotically visiting the nearest stellar and exoplanetary systems have turned into well-funded programs with detailed plans. Meanwhile, our radio and optical searches for coherent signals from technologically communicative life forms continue apace. These monitoring efforts have yet to find any reproducible results, but they deserve society's continued support. To these myriad efforts, I say, *carpe posterum* – seize the future!

Recommended reading and resources

Online appendices to this book

The appendices to this book include seasonal maps of the celestial sphere and seasonal sky charts, along with facts and figures on the brightest-appearing stars, the nearest (naked-eye) stars, and prominent star clusters and nebulae. All are available at oneworld-publications.com/work/astronomy/.

General interest

Astronomy by C. J. Peterson, CliffsQuickReview, IDG Books Worldwide (2000). A concise counterpart to the present narrative.

Astronomy by A. Fraknoi, D. Morrison, S. C. Wolff and other contributors (2016–). A free online college-level textbook: openstax.org/details/books/astronomy. Also available in hard copy.

AstronomyNotes by N. Strobel (2001–). A less flashy but well-crafted free online introduction at the college level: www.astronomynotes.com. The first of its kind.

Classifying the Cosmos – How We Can Make Sense of the Celestial Landscape by S. J. Dick, Springer (2019). A clever presentation of all that the cosmos contains – organized into kingdoms, families, and classes akin to our biological classifications.

Galaxies and the Cosmic Frontier by W. H. Waller and P. W. Hodge, Harvard University Press (2003). A comprehensive survey of galaxies in all their fascinating variety and of the larger cosmos which spawned them.

The Milky Way: An Insider's Guide by W. H. Waller, Princeton University Press (2013). A descriptive exploration of the contents, structure, dynamics, and evolution of our home Galaxy.

Origins: Fourteen Billion Years of Cosmic Evolution, by N. D. Tyson and D. Goldsmith, W. W. Norton & Company (2004). A sweeping account of the cosmos through time – from the first particles to galaxies, stars, planets, and life. Companion to the NOVA television mini-series *Origins*.

History

The Sleepwalkers: A History of Man's Changing Vision of the Universe by A. Koestler, Penguin Books (1990). An engrossing story of our cosmic understandings – from the Babylonians to Copernicus, Kepler, Galileo, and Newton. With erudite commentaries on science vs. religion over the ages.

A Short History of Nearly Everything by B. Bryson, Broadway Books (2003). A more contemporary and humorous take on the human side of science, with the history of astronomy featured along with that of physics, chemistry, biology, and paleontology.

Parallax: The Race to Measure the Cosmos by A. W. Hirshfeld, W. H. Freeman (2001). A lively tale of the astronomers who strived to fathom distances to the stars.

The Glass Universe by D. Sobel, Viking/Penguin Random House (2016). A rich profile of the women at Harvard College Observatory who took the spectroscopic measure of the stars, thus founding the field of stellar astrophysics.

Lonely Hearts of the Cosmos by D. Overbye, Harper Collins (1991). A personality-rich history of observational cosmology during the twentieth century.

Magazines

Sky & Telescope: skyandtelescope.org
Astronomy: astronomy.com
Space.com (online magazine)

Other resources

Astronomy Education Clearinghouse: naec-us.org

Astronomy Picture of the Day: apod.nasa.gov/apod/astropix.html

An Atlas of the Universe: www.atlasoftheuniverse.com

Black Hole Encyclopedia – listing of stellar, intermediate-mass, and super-massive black holes: blackholes.stardate.org/objects.html

Cosmic origins of the elements: apod.nasa.gov/apod/ap171024.html

Cosmological Expansion Calculators: ned.ipac.caltech.edu/help/cosmology_calc.html

The Extrasolar Planets Encyclopaedia: exoplanet.eu

The Event Horizon Telescope collaboration: eventhorizontelescope.org

Gaia Mission: sci.esa.int/gaia

Sol Planetary System Data (concerning planetary bodies in our Solar System): www.princeton.edu/~willman/planetary_systems/Sol

Teachable Moments in Astronomy and Astrophysics: https://sites.google.com/site/sciencegazette/home/teachable-moments-in-astronomy-astrophysics

Technical Notes, etc. from *Galaxies and the Cosmic Frontier*: cosmos.phy.tufts.edu/cosmicfrontier/main.html

Visualizations of transient X-ray emission from neutron stars and black holes: xte.mit.edu (see "ASM The Movie" page)

Zooniverse citizen science platform: www.zooniverse.org

Index

References to images are in *italics*.

aberration of starlight 20–1
AdvancedVirgo detector 223
airplanes 21
Alnilam 11
Alnitak 11
Alpha Centauri A 62–3, *80*, 105,
        106
    and brightness 108, 109
    and color 109–10
    and spectral classification 111
Alpha Centauri B 63, 105, 124
Alpha Orionis, *see* Betelgeuse
Anaximander 42
ancient Egypt 9–10, 23–4, 42
ancient Greece 10–11, 42–4, 45–7
Andromeda Galaxy (M31) 17, 71,
        151, *152*, 164
Antarctica 19
Antares 12, 63
Antlia the Air Pump 12
Aquila 142
Arab world 11
Arabic-named stars 10–11
archeoastronomy 9, 12
Arcturus 10
Aries the Ram 23–4
Aristarchus of Samos 45–6
Aristotle 43
asteroid belt *80*, 93
astrology 23–4

Astronomia Nova (New
        Astronomy) (Kepler) 51
Atacama Large Millimeter
        Array 199–200
atmosphere 35–6, 88, 196, 234, 238
    and circulation 20, 83–4
    and Earth 95, 97
    and Neptune *91*, 92
    and Saturn 85, 87, 236
    and stellar 110, 136, 140, 207,
        213, 217
    and Sun 98, 101, 102, 107
atomic epoch 183–4, 185
atoms 100–1, 233–4
Auriga the Charioteer 131

Barnard, E. E. 196, 197
Barnard's Star 108, 109, 110, 111
Bayer, Johann: *Uranometria* 12
Bell, Jocelyn 215
Bessel, Friedrich 61
Beta Orionis A, *see* Rigel
Betelgeuse 10, 11, 135, 136–7
big bang, *see* hot big bang
binary star systems 117–18
al-Biruni, Abu Rayhan 11
Black Eye Galaxy 154
black holes 192, 211, 217–18, 223,
        238
    and event horizons 218–19, 238

and intermediate-mass 220
and stellar 218–19
and supermassive 221–2
blueshift 164
Bode's Galaxy (M81) 159
Bradley, James 61
brightness 107–8
brown dwarfs 113
Bruno, Giordano 40, 50
bulges (galaxian structure) 127,
129, 130

calcium hydroxide 110
Callisto 85
Canes Venatici 155
Canis Major 62
Canis Minor, see Sirius
Capella 10
carbon 145, 207–8, 209, 233–4, 236
and neutron stars 215
and white dwarfs 212
carbon dioxide 84, 92
carbon monoxide 142, 143, 197
Carina Nebula 17, 144
Carina the Keel 141
Cassiopeia A 217
Cavendish, Henry 56
celestial equator 21–4
Cepheid variables 67, 70–2, 139,
140–1
Cepheus 198
Ceres 93
Chandrasekhar, Subrahmanyan 213
China 12
chromosphere 101–2
Cigar Galaxy (M82) 158–9
Clement VIII, Pope 50
clouds 64, 67, 143, 145, 146; see
also dark nebulae; Eagle

Nebula; gaseous nebulae;
Large Magellanic Cloud;
Small Magellanic Cloud
clusters 131–5
Coalsack 142
cold dark matter 228
collapse, see gravitational collapse
color–magnitude diagrams
(CMDs) 133, 134
colors 109–10, 112, 114, 206
and spiral galaxies 154, 156
and star clusters 133, 134
Columbus, Christopher 44–5
constellations 10–12, 13, 17
Copernicus, Nicolaus 47–8, 49–50
Coriolis effect 20
corona 101–2, 103
coronal mass ejections
(CMEs) 102–3
cosmic microwave background
(CMB) 174–9, 181, 184, 185
cosmos 40–1
and debris 93–4
and expansion 164–7, 184, 230–2
and genesis 173–9
and human beings 233–4
and zooms 75–6
COSMOS Redshift7 190
Crux 142
Cygnus the Swan 123, 125, 142,
145–6

Dark Age epoch 186, 187
dark energy 5, 211–12, 229–32, 237
dark matter 5, 69, 130, 211–12,
224–6, 237
and dark energy 230, 231
and MOND 228–9
and ordinary 226–7

and simulations 187–9
dark nebulae 196–200
De revolutionibus orbium coelestium
    (On the Revolutions of
    the Heavenly Spheres)
    (Copernicus) 50
declination 21, 23
Deneb 10
density 85, 88, 96–8, 156–7, 212
deuterium 151, 182, 183, 184, 226
distance 12, 16, 29, 35–6, 39, 41–2,
    75–6, 80
    and Earth 43–4, 46–7, 49
    and galaxies 164–7, 173
    and Gould's Belt 64–5
    and Jupiter 84–5
    and Milky Way 67–71
    and star clusters 133
    and stellar 105, 106–7
    and Sun 52, 54–7, 58–60, 61–2
    and super stars 136
    and Uranus 88
    and Virgo Supercluster 71–3
Doppler shift 118, 123, 136, 176,
    219
    and galaxies 161, 164
Double Cluster 17
Draco the dragon 25

Eagle Nebula (M16) 144, 198
Eames, Charles and Ray 76
Earth 15–17, 42–5, 195, 234–5,
    238–9
    and circumference of 43–45
    and cosmic debris 94
    and migration of the night
      sky 26–31
    and Moon 31–5
    and night sky 26–8

and Solar System 46, 48, 49–50,
    56–7, 58, 79–80, 82–3
    and spin axis (precession) 19–23,
    24–6, 36–8
eclipses 33–6
Einstein, Albert 98, 121, 220, 222,
    230
electrons 100, 110, *148*, 149,
    182–4, 208
    and black holes 218
    and dark matter 226, 228
    and gaseous nebulae 143–4
    and neutron stars 215
    and nuclear epoch 183
    and white dwarfs 212
Enceladus 87
energy 5, 85–6, 92
    and Sun 95, 97, 98, 99–102
    see also dark energy
Enlightenment, the 54
epochs 26, 179–84
equator 17, *18*, 20, 21–5, 45, 200
Eratosthenes 43–5, 46, 47
Eta Carinae 141, 144
ethane 236
Eudoxus 11
Europa 85, *86*
European Gravitational
    Observatory 223
Event Horizon Telescope
    (EHT) 222
event horizons, *see* black holes
exoplanets 122–5, 237
explosions 214
extraterrestrial life 235–7, 238, 239

al-Farghani, Ahmad 11
Ferdinand, King 45
Flamsteed, John: *Atlas Celeste 13*

Ford, Kent 224
Fornax the Furnace 12
Fornax–Eridanus supercluster 74
Foucault's pendulum 19–20

Gaia mission 137
Gaia satellite 64
Galactic Center Group
    (UCLA) 221–2
galaxies 17, 71, 150–1, 152–3,
    186–7
  and active nuclei 159–62
  and cosmic expansion 164–7
  and elliptical 157–8
  and giant spiral 154–7
  and Hubble's classification 153–4
  and Local Group 151–2
  and merging 190–3
  and questions 237–8
  and redshift 173
  and reionization 189–90
  and simulated scenarios 187–9
  and starburst 158–9, *160*
  see also Milky Way Galaxy
Galilean moons 85, *86*
Galileo Galilei 52–4, 59, 60, 127,
    132
Gamow, George 174
Ganymede 85
Garnet Star 63
gas giants, *see* Jupiter; Saturn
gaseous nebulae 65–6, 68–9, 141–7
Geb *10*
Gemini the Twins 131
Genzel, Reinhard 221
geometric parallax 59–61, 64
Ghez, Andrea 221
giant elliptical galaxies 157–8

giant spiral galaxies 154–7
Gould, Benjamin 63
Gould's Belt 63–6
gravitational collapse 200–2
gravitational radiation 222–4
gravitational waves 211, 222–3
gravity 54–6, 117, 130, 200–2
  and Jupiter 84–5
  and neutron stars 215–16, 217
Great Nebula 17
Great Rift 142

h and chi Persei 131, *134*
Halley, Edmond 54, 55
halos 35, 65, 127, 130, 135, 146–7
  and Galactic *66*, 67, 69
Harmonices Mundi (Harmony of the
    World) (Kepler) 51–2
Haro, Guillermo 200
helium 83, 85, 92, 98–9, 189–90
  and gaseous nebulae 142
  and hot big bang 184
  and Sun 100
  and Sun-like stars 205, 207
hemispheres 17, 19–20, 25–6, 36–8
Herbig, George 200
Herbig–Haro objects 200
Herschel, John 63
Herschel, William 61
Hertzsprung, Ejnar 71, 113
Hertzsprung–Russell
    diagrams 113–16, *138*, *139*,
    140, *206*
Hewish, Anthony 215
hieroglyphics 9
high-mass stars 209–10
high-velocity clouds (HVCs) 146
HII regions 143–5

Hipparchus 11, 23, 107
Hooke, Robert 61
Hooker Telescope 153, 164
Horologium the Pendulum
     Clock 12
Horsehead Nebula (Barnard 33) 197
hot big bang 5, 151, 171–85, 174,
     226–7
hot dark matter 227
hot gas 65–6, 96–7
Hoyle, Sir Fred 174
H–R, see Hertzsprung–Russell
     diagram
Hubble, Edwin 72, 153–4, 164,
     165–7, 174
Hubble constant 165–6, 174
Hubble Space Telescope (HST) 72,
     163, 186, 229
   and dark matter 224–5
   and dark nebulae 198–9
   and gaseous nebulae 144
   and Neptune 92
   and spiral galaxies 154
Hubble Time 167, 174
human beings 233–4, 237
Humason, Milton 164
Hyades 12, 17, 131, 133, 134
hydrogen 85, 92, 184, 189–90
   and gaseous nebulae 142,
     143–4, 146
   and the Sun 100
   and Sun-like stars 205, 207
   and Uranus 88–9
   and white dwarfs 214
hydrostatic equilibrium 96
Hygiea 93

IC 10: 71

ice giants, see Neptune; Pluto; Uranus
inflationary epoch 180–1
infrared 144, 149, 160–1, 184, 198,
     236
intermediate-mass black holes 220
intermediate-mass stars 208, 209,
     212
intermediate-velocity clouds
     (IVCs) 146
International Astronomical
     Union 4, 12
International Space Station (ISS) 16
Io 85
ions 100, 144; see also reionization
     epoch
Isabella, Queen 45
Iran 11

James Webb Space Telescope 125,
     w163, 186
Jet Propulsion Laboratory (JPL) 90
Jupiter 28, 29–31, 48, 52, 53, 58
   and Solar System 80, 83, 84–6
   and viewing 81, 82

Kepler, Johannes 50–2, 55, 117, 118
Kepler Space Telescope 123–4, 125
Kuiper belt 80, 93

La Caille, Nicholas Louis de: Star
     Catalogue of the Southern
     Sky 12
Lambda Cold Dark Matter
     (λCDM) 188, 189, 229
Laniakea 41, 74
Laplace, Pierre-Simon 200
Large Magellanic Cloud
     (LMC) 69–71, 151, 152

Laser Interferometer Gravitation-Wave Observatory (LIGO) 223
latitude 17, *18*, 19, 20–3, 45
Leavitt, Henrietta 67, 70–1, 141
Lemaître, Georges 72
life forms 235–7, 238, 239
limiting horizons 15–16
lithium 182–3, 184
Local Bubble 65, *66*
Local Group 5, 69–71, 72, 151–2
longitude 21–3
lookback time 172–3, 191–2, 230
low-mass stars 203–4
luminosity 108–9, 113, *114*, 115, 119–22
    and Sun-like stars *206*, 207
    and super stars 136–7
    and variable stars 139–41
lunar eclipses 33–6
Lyra the Harp 123, 125

M77 galaxy 161
M84 galaxy 157
M87 galaxy 157, 158
M94 galaxy 155
M95 galaxy 155–6
Magellan, Ferdinand 70
Magellanic Clouds, *see* Large Magellanic Cloud (LMC); Small Magellanic Cloud (SMC)
magma 83
magnetic fields *89*, 90, 91, 216–17
    and Sun 101–2, 103–4
Maragha observatory 11
Mars 28, 29, *30*, 48, 52
    and Kepler 50, 51

    and Solar System 83, 84
    and viewing 81–2
masses 116–19
    and black holes 220–2
    and high-mass stars 209–10
    and intermediate-mass stars *208*, 209, 212
    and low-mass stars 203–4
    and luminosity 119–22
    and white dwarfs 213
Massive Compact Halo Objects (MACHOs) 226
matter epoch 183
Maunder Minimum 104
Mercury 28–9, 81, 82–3, 84
Mesopotamia 11
Messier, Charles 131
Messier 87 galaxy 222
Messier 109 galaxy *129*, 130
meteorites 94, 195
methane gas 88, 92, 236
Microscopium the Microscope 12
Middle East 11
Milky Way Galaxy 5, 67–9, 126–7, 195
    and black holes 221–2
    and dark heart 147–9
    and delineation 127, *128*, 129–30
    and dwarf galaxies 71
    and gaseous nebulae 141–7
    and Gould's Belt 64–5
    and Hubble classification 153
    and star clusters 131–5
    and super stars 135–7
    and variable stars 137–41
    and Virgo Supercluster 72
Mintaka 11

Mira variables 139
Mirphak (Alpha Persei) 63
Modified Newtonian Dynamics
      (MOND) 228–9
Moon, the 16, 19, 31–3, 195
   and eclipses 33–6
   and Solar System 46–7, 49,
      82–3, 84

NASA 90
National Mall (Washington,
      D.C.) 79–80
Native Americans 12
NEOWISE Comet 82
Neptune 28, 58, 80, 82, 90–2
neutron stars 211, 214–17
New Horizons mission 92–3
Newton, Isaac 54–6, 228–9
night sky 26–8
North Pole 17, *18*
nuclear activity in galaxies 159–62
nuclear epoch 182–3
nucleosynthesis 226–7
Nut (Egyptian goddess) 9–10

observatories, Middle Ages 11
Olbers' Paradox 172, 184
Oort Cloud *80*, 84, 93
Ophiuchus 108, 110, 142
ordinary dark matter 226–7
Orion Spur 65–7
Orion the Hunter 11, *13*, 17, 19
   and celestial equator 21, 23
   and dark nebulae 198, 199
   and gaseous nebulae 144
   and Gould's Belt 63, 65
   and super stars 135
oxygen 144, 145, 207–8, 209, 236

and source 234
and white dwarfs 212
ozone 236

Pallas 93
parallax effect 59–61, 64
Parmenides 42–3
particle epoch 182
Pauli exclusion principle 212
Pavo–Indus supercluster 74
Penzias, Arno 175
Perseus the Hero 17, 63, 131
Perseus–Pisces supercluster 74
photons 100–1, 147–8, 183
photosphere 101–2
Pinwheel Galaxy (M33/101) 71,
      151, 154, 155
Pisces the Fish 24
Planck epoch 179–80
planetary motion 50–5
planetary nebulae 143, 145
plasma 100, 103–4
Plato 43
Pleiades 12, 17, 131, *132*, 133, *134*
Pluto 58, 92–3
Polaris 17, 21, 25
Powers of Ten 75–6
precession 24–6
*Principia Mathematica Philosophiae
      Naturalis* (Newton) 54
Procyon 105
protons 98–9
protoplanetary disks 198–200
protostars 200–2
Proxima Centauri 124, 202
Ptolemy, Claudius: *Almagest* 11,
      47–8
pulsars 211, 215–17

quantum tunneling 98–9
quasars 161–2, 192
Quechua Indians 142

Ra 9
radiation
    and CMB 174–9, 181, 184, 185
    and gravitational 222–4
    and the Sun 100–1, 102–3
    and Sun-like stars 204–5, 208
    and Uranus 88
    and variable stars 140
redshift 162, 164–5, 173, 190–2
reionization epoch 189–90
religion 50
Rigel 11, 63, 135
right ascension (RA) 23
ring systems 35, 82, 84, 85, 87, 156
rock paintings 9
rocks 195, 234–5
Roman Empire 10–11
Rossi X-Ray Timing Explorer 219
rotation (planetary) 20, 154, 156,
    198, 200–2
    and Saturn 85
    and Uranus 89
RR Lyrae variables 67, 138–9, 140
Rubin, Vera 224
Rudolph II, Holy Roman
    Emperor 51
Rudolphine Tables (Kepler) 51
Russell, Henry Norris 113

Sagan, Carl 94, 196
Sagittarius the Archer 24, 25, 67–8,
    144
Samarkand observatory 11
Saturn 28, 48, 52, 58
    and Solar System 83, 85–7

and viewing 81, 82
    see also Titan
Scale of the Universe (graphic) 76
Scorpius the Scorpion 63
seasons 36–8
Seven Sisters, see Pleiades
Seyfert, Carl 160
Shakespeare, William 90
Shapley, Harlow 67–8, 141
shearing motion 154–6
Shu 10
Sirius 10, 12, 62–3, 105, 122
    and binary system 117–18
    and brightness 107, 108, 109,
        116
    and color 110
    and spectral classification 111
61 Cygni 61, 107, 108
Small Magellanic Cloud
    (SMC) 69–71, 141, 151, 152
solar eclipses 33–6
Solar Neighborhood 41, 59, 61–2,
    64, 105–6
    and brightness 107–9
    and colors 109–10
    and distances 106–7
    and exoplanets 122–5
    and Hertzsprung–Russell
        diagram 113–16
    and masses 116–19
    and mass–luminosity–lifetime
        relations 119–22
    and spectral classifications 110–
        13
Solar System 45–57, 58, 59–61,
    79–80
    and black holes 220
    and cosmic debris 93–4
    and formation 195–6, 238

and gas giants 84–7
and ice giants 88–93
and observing 81–2
and Sirius 62–3
and terrestrial planets 82–3
see also Sun
Sombrero Galaxy 154
Soul Nebula (W5) 198
Southern Cross 17
spectral classifications 110–13
Spica 10
spin axis 87, 89–90, 91, 196,
    216–17
and Earth 21–2, 24–5, 36, 38
Spitzer Space Telescope 163, 198
Square Kilometer Array 186–7
starburst galaxies 158–9, *160*
stars
    and clusters 131–5
    and formation 190–3, 194–5,
        196–202
    and high-mass 209–10
    and intermediate-mass *208*,
        209, 212
    and lifetimes 119–22
    and low-mass 203–4
    and Sun-like 204–5, *206*, 207–8
    and super 135–7
    and variable 137–41
    see also neutron stars; white
        dwarfs
Steady State theory 174
stellar black holes 218–19
stellar parallax 106
Stonehenge 9
summer solstice 25
Sun, the 5, 95–8, 195, 238
    and astrology 23–4

and clusters 134–5
and color 109, 110
and energy transport 99–102
and Gould's Belt 64
and lifetime 120–1
and luminosity 137
and Moon 33–5
and orbit 27–8
and planets 28–9
and power 98–9
and rotation 201
and seasons 36–8
and solar activity 102–4
and viewing 81
see also Solar System
Sun-like stars 204–5, *206*, 207–8
Sunflower Galaxy (M63) 154, 155
sunspots 102, 104
super-Earths 125
super stars 135–7
supermassive black holes 221–2
supernovae 214
    and remnants 143, 145–6, 210

T Tauri variables 140
Tau Ceti 107, 108
Taurus the Bull 12, *13*, 17, 131,
    145, 198
temperature 64, 84, 89, 110–13,
    120
    and star luminosity 113, 115
    and the Sun 96–8, 104
    and Sun-like stars 204–5
TESS (Transiting Exoplanet Survey
    Satellite) 125
thorium 234–5
3C 273 galaxy 162
Thuban 25

tilt, *see* spin axis
Titan 87, 92, 236
titanium oxide 110
Triangulum Galaxy (M33) 71, 151,
    152, 164
Triton 91, 92
al-Tusi, Nasir al-Din 49
Tycho Brahe 50, 51, 60

universal expansion, law of 72–3
Universe, the 171–3, 184–5
    and epochs 179–84
    and genesis 173–9
    see also cosmos; galaxies
uranium 234–5
Uranus 28, 58, 82, 88–90
Ursa Major 12, 42
Ursa Minor 17, 25

variable stars 137–41
Vega 10, 25
Vela 146
Venus 28, 29, 53–4, 56, *57*
    and observing 81
    and Solar System 82–3, 84

vernal equinox 22–3
Vesta 93
Virgo Cluster 71–2
Virgo detector, *see* Advanced Virgo
    detector
Virgo Supercluster 72–3, 152
Virgo–Centaurus–Hydra filament
    of superclusters 73–4
Voyager 2: 89–91, 92

Uzbekistan 11

water 85, 88, 92, 125, 234, 235
Weakly Interacting Massive
    Particles (WIMPs) 227, 228
Whirlpool Galaxy (M51) 154
white dwarfs *114*, 115, 146, 207–8,
    211, 212–14
Wilson, Robert 175
winter solstice 25
wobbles 26, 124–5

Zwicky, Fritz 224